D0088859

Fifteen Candles

395.24
FIF

Farmers Branch Manske Library
13613 Webb Chapel
Farmers Branch, Tx 75234-3756

rayo *An Imprint of* HarperCollins*Publishers*

Fifteen Candles

15 Tales of Taffeta,
Hairspray, Drunk Uncles,
and other
Quinceañera Stories

an anthology edited by

Adriana Lopez

FIFTEEN CANDLES. Copyright © 2007 by ADRIANA LOPEZ. All rights reserved. Printed in the United States of America. No part of this book may be used or reproduced in any manner whatsoever without written permission except in the case of brief quotations embodied in critical articles and reviews. For information, address HarperCollins Publishers, 10 East 53rd Street, New York, NY 10022.

HarperCollins books may be purchased for educational, business, or sales promotional use. For information, please write: Special Markets Department, HarperCollins Publishers, 10 East 53rd Street, New York, NY 10022.

FIRST EDITION

Designed by Janet M. Evans

Library of Congress Cataloging-in-Publication Data has been applied for.

ISBN: 978-0-06-124192-5
ISBN-10: 0-06-124192-X

07 08 09 10 11   DIX/RRD   10 9 8 7 6 5 4 3 2 1

Table of Contents

Reluctant Damas & Chambelanes

The Dreamers

For my mother and her fortitude,
when it seemed everyone
in my peer group
was considering a jump off a bridge . . .

An Introduction

I can't remember my fifteenth birthday. I was probably punished as usual, locked up in my Laura Ashley—wallpapered room, sobbing to my Culture Club poster, asking: "Why me? *Why?*" Life's emotional arc seemed so extreme back then. Like a poorly developed plot to an MTV music video or an after-school special whose storyline was equally as calamitous as my own life's.

It's all a lot of information to take in for a new adult brain. Circa XV, you're basically a fully formed person conscious of your surroundings. A bit precocious? Yes. Strange? Absolutely. But nonetheless, you still have a pretty clean slate to be proud of. You're ripe with ideas about the world and you're at your physical prime. Yet despite these attributes, you're absolutely helpless, *useless*, for two fundamental reasons: 1) your flaming hormones won't let you think straight, and 2) you're still a citizen of the republic of the family home. You're basically stuck giving in or defecting.

On the edge of adulthood, turning fifteen can pose many existential questions about your future role in society. Especially when the deep-seeded symbolism of the quinceañera ritual is involved. This ultimate coming-out party for Latinas in the States and in Latin America has always, and will always remain, a rite of passage to cherish forever or relive in therapy first, and eventually grow from (most of this collection's contributors belong to the latter group).

This ostentatious girl-to-woman transformation tradition has been hard to shake for Latinos since its rougher beginnings in

Aztec culture, where fifteen-year-old girls were feted for being ready for marriage and procreation, and boys for war. After Europe spread its influence over the Americas, quinceañera parties for upper-crust debutantes had more of a cotillion bent to them. This formal introduction of rosy-cheeked girls to society included the not-so-Latin-American waltz and usually came with a crown, a fancy gown, and fifteen dancing couples.

In the last decades, having a quinceañera party has gone in and out of fashion like any gyrating teen heartthrob of the day. But in the last few years, cities all across the United States have witnessed quite the quinces comeback. Call it a case of major Latin pride, or Bat Mitzvah and Sweet Sixteen envy, but flocks of Latinos are resurrecting this ritual to all-new millennium heights with ceremonies as opulent as a *Dynasty* wedding. Pyrotechnics aside, we're still a traditional people and most of the quinceañeras' earmarked traditions still linger, albeit with new twists for our multiculti times. Some families still opt for modest backyard parties, some do it up in big catering hall parties, some rent out Disney World, and some prefer a reflective religious ceremony. But all include the girl, her family and friends, the big expectations, the nerves, and eventually the messy mush of memories.

It doesn't matter whether you were the only Latino family in town or if your neighborhood grocery had a fully stocked Goya foods section, if you lived in the States, a quinceañera party affirmed your Latinoness. And at fifteen, many writers from this collection weren't sure about this multigenerational-Latino-Catholic-formal-dance-party-with-free-alcohol thingamajig. It could be kind of surreal for the average young Latino. Maybe you'd been to a wedding or two in your youth, but this was about celebrating a peer's initiation into the trappings of adulthood. No groom in sight, just God. *"I mean, gosh!"* the average preteen might ponder, *"there's this holy reception first, then we're at this huge party, she's suddenly wearing makeup and jewelry, she's waltzing, she's followed around by girls in matching dresses and dudes in tuxes, her father dances with her and then encourages her to dance with other guys, and she's trading in her flat slippers for sexy heels?!"*

Whether you're related to the quinceañera, dating her, in love with her mother, a friend of a friend of hers, or just there for the free food, you're still at this function asking yourself questions about where you fit into this spectacle. Under the heel-beaten floorboards of this seemingly harmless fiesta lie bigger, thornier issues for the preteen Latino to consider: the curiosities of cultural traditions, recurring acne, the church, appropriate dancing techniques, peculiar relatives, covert drinking, unflattering formal attire, the opposite sex, the stereotypes and mandates of your own sex and, most importantly, whether this means you're now allowed to have sex.

In my search for fifteen writers with a personal story to tell about a quinceañera event in their lives, I found the majority of tales fell into five types of categories: the writer who had a party, the writer who was forced to have a party, the writer who dreamed about a party, the writer who was serendipitously invited to or crashed a party, and the writers asked to be a *dama* (pseudo-bridesmaid) or a *chambelán* (quasi-groomsman) in the quinceañera's court (adolescent bridal party). I was thankful for the bounty of story submission ideas I received, and was relieved that there were plenty out there with a fun-loving sense of humor willing to be frank about the freaky stuff that had happened to them on the way to Quincelandia.

Aiming to steer clear of monochromatic depictions of the quinceañera event, I'm proud to say that *Fifteen Candles'* contributors are as motley as that '80s rocker "crue." They vary by gender, orientation, social class, region of the Americas, and generation. And though they were chosen for their originality in voice, experience, and point of view, in the end, their stories are all surreptitiously linked by humor, sadness, and a lot of self-discovery.

For those looking for a how-to book for planning a quince-añera bash, please put this book down; this was not intended for you. The parties depicted on these pages aren't always that pretty. *Fifteen Candles* aims to poke fun at the ritual, question it, re-imagine it, challenge it, and see it through the eyes of a newcomer. Divided into five thematic sections, these are candid snapshots of fifteen writers looking back on their former, younger selves, trying to understand what it meant.

For some, such as in the opening section, The Romantics, a hormonal high seemed to have propelled them through their experience. Writer Angie Cruz relives a shrouded first kiss with a *chambelán* so cute, he could have been in Menudo, while actor and lawyer Alberto Rosas cursed with looking like Antonio Banderas, accidentally seduces both quinceañera and her mother. And teacher and writer Constanza Jaramillo-Cathcart considers the reason her dashing, white-suited date to a Bogotá quince never called again might have had something to do with his family's ties to the mob.

But of course there were others who had no time for flirting with all the responsibility they had to deal with. In the second section, Queens for a Day, these are the writers who had to grin and bear it as the quinceañera in headlights, ahem, spotlights. Resurrecting uncensored pages from her lock-and-key diary, journalist Fabiola Santiago reflects on her humble party and compares it to how her three very different daughters chose to turn fifteen themselves. Editor Leila Cobo-Hanlon knew from the get-go her dress *had* to be emerald green or she would go naked. Finding that color was tough, but convincing the cool kids from school to attend her house party was even tougher. Teacher and poet Nanette Guadiano-Campos's family made her do it. As a new kid in a not-so-Latino part of Texas, a quinceañera party was the last thing she wanted—especially with Tía Chucha clad in clingy leopard print to mortify her even more.

Then there were those in The Party Crashers section with nothing better to do on a Saturday night than to drop in and cause trouble. Writer Malín Alegría-Ramírez becomes a superhero when she saves a friend's gift from the wrong quinceañera party on a night two rival *taquerías* are holding celebrations in the same hall. Performer and writer Adelina Anthony drops by the bad girl in school's church quince for free enchiladas. But when she spots the quinceañera's divine mother, it's not just her classmate's coming-out party anymore. Writer and filmmaker Eric Taylor-Aragón just couldn't say no to his earnest coworker Antón's invitation. So when he attends the quince of a girl he never met, the strange party leaves him, literally, in shreds by the morning—but with an apotheosis on life.

And have you ever been invited to a party you didn't really want to go to, but felt you *had* to? For the three writers in the section Reluctant Damas & Chambelanes, wearing the frills and jackets were angst-provoking for personal reasons of their own. Writer Michael Jaime Becerra documents a relative's numerous experiences as an in-demand *chambelán* and the impact these ceremonies had on his relationships with his parents, his sisters, and women in general. When writer Barbara Ferrer was only five years old, she was thrown into a junior court as a miniature *dama*. She hasn't recovered from the psychological consequences of the arduous beauty preparations since. And performance artist Monica Palacios clearly remembers preferring her purple bell bottoms, her Jane Fonda hairdo from *Klute*, and her basketball uniform over that itchy-rash-causing *dama* dress that her good friend wanted her to wear.

While some hemmed and hawed over having to attend yet

another fifteen-candled episode, others sat and reflected on what might have been if destiny had pointed its arrow in another direction. In the final section, The Dreamers, the quinceañera event for these three writers lives in their fanciful imaginations and bittersweet memories. To heal writer Felicia Luna Lemus's adolescent angst, nothing but black velvet–draped buffet tables, bloodred punch, and a grand entrance to Chopin's "Funeral March" in a black silk Queen Victoria mourning gown would do for the accoutrements to her Goth quince. For writer Berta Platas, images of scandalous minuets and big-skirted Givenchy ballroom gowns from 1950s Havana danced merrily in her head. That is, until her parents dropped the bomb that they couldn't afford a party.

The final story in this collection departs from the thematic, lighthearted folly of a young life beginning to the tragedy of a young life taken all too soon. When writer Erasmo Guerra journeys back home to dusty South Texas, a family photograph taken for his sister's humble quinceañera celebration gives way to a waterfall of painful memories surrounding the mystery of his sister's death. In Guerra's words, ". . . she will forever remain a teenage princess, dead before she was able to fulfill her promise as a woman in the world." His touching story puts this tender milestone in a Latin woman's life into sobering perspective.

Maybe, like me, you too find it hard to remember what it was like turning fifteen. All these stories should help bring it all back, push aside those sticky cobwebs, and unveil your hidden drama queen in taffeta. They will remind you of that fantastically ridiculous but sentimentally meaningful time gone by. So go ahead, blow out the candles and make a wish.

Fifteen Candles

The Romantics

El Quinceañero

BY Alberto
Rosas

A few months ago my cousin Ilene asked if I wanted to be a godfather at her quinceañera. This meant I was getting old. Less than a decade earlier, girls had asked me to be their *chambelán* or *chambelán de honor*. At twenty-five, I still considered myself a good candidate for *chambelán de honor*. Ilene disagreed. Although I knew my answer to Ilene would be no, not wanting to hurt her feelings, I told her I would get back to her.

Being in a quinceañera again went against my unwritten rule. I had called it quits after Yvette's party in 1997.

After Ilene left my apartment, I studied my aging reflection. Hair: black, full, healthy, and gray-free. Wrinkles: none, with the exception of a few character creases on my forehead. Still young. Still *chambelán* material.

My reflection stared at me as I washed my face. The cold water felt fresh against my skin. The water dripped from my face onto the sink. *Droplets of champagne dripped from Yvette's hair as we danced. Small pieces of flan stuck to her face and neck. The melody of the waltz blared through the speakers as our bodies moved across the dance floor.*

I couldn't shake the memories. Images from that quince played in my mind like a movie. I saw her face, how sweet she looked in her flowing white gown. Those innocent green eyes stared out from a ghostly white face. Her mother's face appeared: seductive green eyes contrasting beautifully with naturally tanned skin. Then the flan came to mind. It was imported from Tijuana, the *padrino* said, and it was the best flan I had ever tasted.

Yvette stood in her backyard and barked orders at the group of *damas* and *chambelanes*. The majority of the group arrived late, and none of them knew the steps to the waltz.

"You all need to do it good," Yvette ordered.

"Maybe if we danced to hip-hop," a *dama* said.

"This is gay," a *chambelán* said.

We turned to look at the gay waltz coordinator, Esteban, who waved one hand in the air and mouthed "Whatever." We also looked at Yvette's lesbian mother, Ingrid, who just shook her head and said nothing.

Yvette took my hand. Esteban pressed a button on the oversized boom box and Chayanne's "*Tiempo de Vals*" began to play.

Chayanne's waltz was a smooth mixture of romantic pop song with a waltz's repetitive three-count bass. This seemed to be a popular song for quinceañeras; it was the third time I had danced to it.

The couples danced around the crowded backyard, minimizing their moves to accommodate the small space. Yvette stared at her feet as she danced.

"Look at me," I said.

This was my eighth or ninth time as a *chambelán*, my fourth as the *chambelán de honor*. By seventeen, I was a waltz aficionado.

At my first quinceañera, when I was eleven years old, I couldn't decide whether I had two left or two right feet. The mariachi ballads and *banda* polkas were too complicated for my coordination. Besides, I hated that mariachi and *banda* shit. It wasn't until a few years later, when I was introduced to salsa and merengue, that I had discovered my hidden dancing abilities.

"You're a pretty good dancer," Yvette said.

"I'm the Latino John Travolta."

Though Yvette had been a *dama* in various quinces, the added stress of being *the* quinceañera turned her into a virgin on the dance floor. Plus, waltzing wasn't her thing; it was just something that she *had* to do. The waltz was interconnected with the quince and it was impossible to have one without the other. It would be like having beans without rice, a *carne asada* barbecue without beer, or a piñata without candy. It would be like having a daughter but not throwing her a quinceañera party. It was tradition.

Both the *chambelanes* and *damas* slouched as they danced, their shoulders hunched forward. The boys danced left to right in a simple one-two count instead of the one-two-three count.

"Guys, please," Esteban said, "don't drag your feet like you got sandals on or something."

"I hate this waltz shit," someone said.

During the break, we ran to the cooler for sodas. The seven *damas* stood in one corner of the backyard. Six of the chambelanes stood about on the front sidewalk, while Chuy went inside to watch TV. My slacks and polo shirts clashed with their baggy jeans and T-shirts. Feeling out of place, I remained in the yard and sat near Esteban and Ingrid Garcia, Yvette's mother.

Yvette stood next to a husky boy about my age. He had a few thick whiskers outlining a thin goatee. I envied his thick facial hair compared to my sprouting peach fuzz. I met him a year earlier around the time I met Yvette, when she and I were *dama* and *chambelán* in Alma's quince. Yvette and I became friends after that and I always thought of her as a kid sister. It was at Alma's quince that I met Carlos, Yvette's boyfriend. He arrived at the quince with baggy pants and a bandana hanging from his rear pocket. Alma told him to leave, Alma and Yvette got into an argument, and the two never spoke again. So when Yvette asked me to be her *chambelán de honor*, I asked, "What about your boyfriend?"

"He can't dance."

"Maybe I can teach him some moves."

I didn't know any of Yvette's friends. The seven *chambelanes* appeared to be young gangbangers or wannabes. The film *Mi Familia* had been released about a year earlier, and the *chambelanes* aspired to be *vatos locos* like Jimmy Smits's character. Their short-sleeved shirts and tank tops revealed the latest trends in homemade tattoos, which consisted of skeletal outlines of partial images and misspellings, like the *chambelán*'s tattoo that read "Yes Sí" instead of "Jesse." The *damas* wore too much makeup and talked nonstop about Enrique Iglesias, who had recently made it onto the music scene. The *damas* wore Enrique T-shirts and imitated his Spanish lisp.

So I sat alone on a bench with the two homosexuals—Esteban

and Ms. Garcia—listening to them talk about shoes, dieting, and telenovelas.

"You know, you look like Antonio Banderas," Ms. Garcia told me.

It wasn't the first time I heard that comparison. The film *Desperado* had been released two years earlier, and every so often someone would make the comment. Antonio and I had the same complexion, the same eyes, and I also had long hair, though not long enough for a ponytail. My black friends started calling me Tony Flags.

I just smiled and nodded toward Ms. Garcia. I seldom spoke with her or with Esteban because I felt uncomfortable with their chosen lifestyles. After all, I had been raised within the macho homophobia of Mexican culture.

Ms. Garcia's full lips twisted into a cocky smile.

Yvette's apartment was a modest two-bedroom in a two-story apartment building located in a middle-class neighborhood in Oakland, California. Their unit rested above the garage and they shared a small common backyard with their other neighbors. Every time I went over there you'd hear some *banda* or salsa tune filling the hot summer air. There were family photographs on the shelves and Aztec artwork covering the walls. Some of the older photos contained a light-skinned man with a thick handlebar mustache. He appeared in many of the photos with Yvette as a toddler: carrying her on his shoulders, posing inside his blue Mustang. He was a good-looking man who never smiled in any of the photos. And as Yvette approached her preschool and kindergarten years, this man just disappeared from the pictures.

Ingrid Garcia had been eighteen years old and pregnant with Yvette when she married this man, and he brought her over from

Mexico. He worked in shipping and receiving and constantly traveled. Life was difficult for Ingrid in the beginning, especially in dealing with her husband's traveling and infidelity. When the police came knocking at her door one day, Ingrid realized that what he'd been shipping and receiving was cocaine. Yvette was five when her dad was sentenced to a federal correctional institution. When Ingrid mailed him the divorce papers, she also decided to never to be with another man again. He had been her first.

After only a short time of frequenting the nightclubs and bars in and around San Francisco, Ingrid reinvented herself. The soft-spoken and shy Mexican girl blossomed into a hard-drinking, hard-partying lesbian. Her husband was imprisoned for eleven years, and in that time she never slept with another man. Until me.

"Can you teach me to waltz?" Chuy asked. At thirteen, Chuy was the youngest *chambelán*. He looked like a walking skeleton in his baggy pants and T-shirt, his belt hanging to his calves. Though he looked like a young *cholo*, he didn't act like one. Probably just pretending to be tough in order to survive those mean streets.

"I keep stepping on Juanita's feet," Chuy said, with a slight crack in his voice. "It's my first quince. I want to do real good."

Chuy reminded me of myself at my first quinceañera. I couldn't dance and was fearful of performing the waltz in front of a crowded hall of people. But I was lucky that the quinceañera and most of the *damas* were my cousins, so they took me under their wing and taught me about dancing the waltz and drinking tequila. Unfortunately for Chuy, he didn't have any cousins in the quinceañera. He was stuck with me.

His aunt lived a few houses from Yvette. We agreed to meet at his aunt's during the week and practice. The plan was to watch Chuy dance alone, pretending he was holding little Juanita in his arms. This proved useless. His frame dropped and he stepped on her imaginary feet. I had no choice but to dance with him.

"Pretend I'm Juanita," I said.

"This is some gay shit."

"Come on, man, this is the only way you'll learn. Male tango dancers in Argentina practice with each other, so that when they meet girls at the dance, they're ready. I saw it on TV."

He slowly dragged his feet across the carpeted room and reluctantly grabbed my hand. "Don't tell anyone about this," he said. "It'll ruin my reputation."

"Mine too."

Chuy danced with me in his aunt's living room. Sometimes I led him when he forgot a step. "You're taller than Juanita," he said.

My girlfriend, Catalina, sat on her couch. She was sixteen, voluptuous, and hadn't had a quinceañera. Her father, who had been chugging Coronas by himself all afternoon, sat on a chair on the opposite end of the living room, watching *Sábado Gigante,* drinking rum and coffee out of a mug, and eyeing me during commercials.

"Shit," Catalina said, as she studied the invitation. "There's a shopping list of godparents." She waved the invitation in my face. "There must be twenty *padrinos!*" She handed me the card. "Everyone can have a quinceañera party if other people pay for it. The quinceañera and her parents don't pay for shit. That's why I didn't have a quince. I didn't want to bother people for money."

Although there was no commercial, her father looked over at us.

"No one loses any money when you think of it," I said. "Take the *padrino de anillo,* for example. He spends three hundred dollars on a ring. He and his family eat about a hundred fifty dollars' worth of food. They drink about a hundred dollars' worth of beer. They listen to free music. Maybe they steal some of the silverware. They come out ahead."

"What about *padrino de vestido*?" her father asked. "That can be one, two thousand dollars."

"Assume the dress is fifteen hundred," I said. "The *padrino* and his family eat a hundred and fifty worth of food. They're pissed that they spent so much money, so they drink three hundred dollars' worth of beer. They take extra food and beer to their car— another two hundred. Maybe *el padrino* gets some bang-bang action in the bathroom. That's got to be worth something. I lost count, but I think *el padrino* made a pretty good investment."

Catalina shook her head. Her father stood up and moved across the living room toward the closet, grabbed his coat, and said he'd be back. All the food talk had probably made him hungry. Or maybe he was going out in search of a good investment of his own.

Catalina and I looked at each other. We heard his footsteps grow faint as he left the house. Keys jangled. A car door opened and slammed shut. An engine coughed to life. A car drove away. Seconds later our clothes were tossed onto the living-room floor and I was on top of her. The couch cushions squeaked. On the TV, Don Francisco wore one of his funny hats and El Chacal played the trumpet as the audience applauded.

Ms. Ingrid Garcia was the second woman I ever slept with. I arrived at Ingrid's apartment like I did every Tuesday and Thursday for practice. The gate to the backyard was wide-open. The rear door to the apartment unit was unlocked. Ingrid stood in her kitchen in a lime green thong and yellow Bart Simpson T-shirt. She held a greenish drink in her hand. My body stiffened and immediately began to perspire.

"No one told you?" she wondered. "Practice was canceled. Esteban couldn't make it." She sensed my discomfort and smiled. "I'm sure you've seen a naked lesbian before." She moved toward me. She placed her drink in my hand. "I'll put something on."

"No, that's okay," I said. "No, I meant . . . I'm leaving."

"Don't be ridiculous," she said. She went to her bedroom, and I sat on the couch.

"You're pretty good at the waltz," she said from the bedroom.

"I've been in a lot of quinces."

"How old are you?"

I was seventeen and four months, but I said, "Eighteen."

"You should stay and have a drink with me." She stood in the doorframe in a pair of tight blue jeans. She reminded me of a taller version of Catalina. I handed her back the drink.

"Taste it," she said. "It's a mojito, a Cuban drink."

"No, thank you."

"Drink it," she said. "How can you know you don't like something if you don't try it?"

I took a sip. "Tastes like toothpaste."

"That's the mint leaves and sugar." She walked to the kitchen and made another. "I first had this in Puerto Vallarta. I used to go there as a kid. It's a beautiful place." She gestured to the drink. "Finish it. I'm making more."

We went outside, sat on the bench, and had more drinks. We smoked two of her Cuban cigars. Her lighter had a picture of a naked woman on it. We talked about the upcoming quince. It was costing her a fortune despite the godparents. She wanted to buy Yvette a car instead, but Yvette had refused.

"A quince is very important," she said. "People compare it to a wedding, but a woman can walk down the aisle more than once. There is only one quinceañera."

After our third mojito and halfway through the cigars, Ingrid asked again, "Has anyone ever told you, you look like Antonio Banderas?" I now realized that she was probably drunk.

I nodded and sipped.

"Yvette says she's a virgin. Is it true?" She stared at me.

"I never touched her!"

"Don't be ridiculous. I know you wouldn't. That's why you're her *chambelán*. I wanted someone I could trust." She asked, "Her little boyfriend, are they sleeping together?"

"I wouldn't know."

She puffed on her cigar and stared off, her eyes not looking at anything, as if her mind had raced back and she imagined herself on those beaches of Puerto Vallarta, a mojito in one hand, a cigar in the other.

"A quince party is a passage from childhood to womanhood. It's a tradition, a celebration of a young girl blossoming into a female sexual being." She finished her green drink in a single gulp. The ice cubes clicked against the glass. "During the waltz, she dances with the main *chambelán*. She is passed to the other *chambelanes*. All of the *chambelanes* and *damas* dance with one another. It's the way of life, everyone fucks everyone. A quince is how the girl announces to the world she is ready to fuck."

I excused myself to the restroom.

As I masturbated, I imagined Ingrid in her green thong, what she looked like under her Bart Simpson T-shirt. Her green thong slowly came off. She lured me onto the bed with her. She kissed me. I kissed her. She ripped my clothes off. Her tan complexion, her firm thighs and legs, her naked body. I was inside of her. She moaned. She exhaled.

She knocked on the door. "Is everything okay?"

Was there a crack in the wall? Had she heard me masturbating? Was there a hidden peep hole in the door? Could she see me with my slacks bunched at my ankles and my hand frozen around my penis?

"Uh . . . What?" I said.

There was silence from behind the door. "Open the door."

"Why?"

"Open the door right now."

"But I . . ."

"I know what you're doing. I can hear you. Open the door."

I pulled my pants up and quickly tied my belt. My hand shook as I opened the door. She stood in the door frame with her arms folded across her chest. She looked at me. I lowered my head in shame. When I glanced up, she was staring at my pants zipper. I turned around and tried to cover myself. Her body pressed against mine from behind. Her warm arms wrapped around me, caressing my chest and stomach, her soft hands working their way downward. "It's been almost ten years since I've held one of these."

I stood in front of my mother's floor-to-ceiling mirrors in a black tuxedo with a pound of gel in my hair. My mom walked into the room and observed me, a proud look on her face. "You look very handsome," she said. She noticed the disarray my black tie was in and quickly removed it. She asked what time the Mass and reception started and my answer was delivered with hesitation. My girlfriend would be there. My thirty-two-year-old lover would be there. The last thing I needed was my parents with their flashing camera.

"I wanted to throw you a quinceañero party when you turned fifteen," she said as she helped with my tie. "Why do the girls get all the fun? Look at the Jewish culture. The boys get a bar mitzvah. The boys celebrate their coming-of-age, their transformation from boyhood to manhood. In a culture as macho as ours, I don't know why we don't celebrate becoming a man just as we celebrate becoming a woman." She had finished with my tie. "Look at you. El quinceañero."

We stood outside the church. Our families and guests sat inside. The *chambelanes* and *damas* practiced their church march. Chuy practiced aloud: "Left, together, right, together, left, together . . ." The *damas* looked great in their light blue gowns. The *chambelanes* wore matching light blue neckties with black tuxedos.

Yvette's sleek white gown complemented her figure. Her body was thin and athletic. She played softball and was on the swim team. But she had this beautifully angelic face stuck on top of that tomboy body of hers. Some of the boys teased her that she looked more like a skinny white girl than a curvy Latina.

Her makeup was subtle, her hair elegantly tied back. It was the most attractive she had ever looked.

The *chambelanes* and *damas* stood in line from the shortest to the tallest couple, with Yvette and me last. The mariachi band played a slow ballad and the first couple walked down the church's center aisle. The church guests rose to their feet.

I thought to myself: *Poor quinceañera only has one party, but as a* chambelán de honor, *this was my fourth quince!* I felt like a movie star walking down the red carpet at a premiere. But I was more like the supporting actor who *thought* he was the star even though everyone couldn't care less about him. I grinned like an idiot. I probably wasn't even framed in the photographs. Perhaps only my right hand appeared in the shot. But I grinned anyway.

Yvette was the star of the movie. As the cameras flashed she couldn't help but smile and cry. Droplets of tears formed in the inner corner of her eyes, but she tightened her jaw and kept smiling, not letting a single tear drip onto her face and damage her makeup job. But her joy got the better of her, and as I looked over at her again I noticed the tears flowing down her cheeks. I squeezed her hand to comfort her and thought about my mother's words: Jewish boys have a bar mitzvah, Latino boys don't have anything.

After escorting Yvette to the front of the church, I took my place in the front pew. From where I sat, I could see Ingrid wearing a skin-tight black dress that clung to her body like a second layer of skin. As the priest spoke, I reminisced about Ingrid. The way she caressed my body, her fingertips brushing against my skin. The way her body pressed against mine in the fitting room, my tuxedo pants lying on the floor. The way the water traveled down her body as we stood in her walk-in shower. As I looked at the priest, the word "fornicate" came to mind.

Catalina sat a few pews behind me. She looked beautiful in her summer dress. We were a year apart, attended the same school, and liked the same things. She was my first love. Although I also enjoyed being with Ingrid, I knew this thing between us would not last. It couldn't. We were different ages and had different interests. The priest lectured about growing up, making decisions, and taking responsibility. By the end of the lecture, I decided to break it off with Ingrid.

We made our way out of the church, Yvette's arm wrapped around mine. I caught a glimmer of a face in the crowd. At first I didn't give that face any thought, with its goatee and its strong brown eyes. I looked again for a few seconds before Yvette and I passed his bench. A younger version of that face appeared in many of Yvette's photographs. But there was something about the man sitting in the pew that did not resemble the man in the pictures. It had nothing to do with age. It had everything to do with time—in this case, prison time—and being changed by it.

On the limousine ride to the reception hall, Chuy and a *chambelán* lit a pair of blunts and within seconds the strong, herbal

stench of marijuana filled the cabin. "Just like the Aztecs," someone said. I didn't have to smoke; there was enough secondhand
smoke lingering inside the cabin. "Even the Aztecs had quinceañeras," someone said. My eyes burned from the thick smoke.
Yvette's large green eyes were glazed over as she carefully grabbed
the joint with her white silk gloves, cautious not to dirty her dress.
She held it as far away from her face as possible, bringing it close
to her only to take the occasional puff. A long cigarette holder
would have gone well with her attire.

Everyone sunk into their seats. Yvette leaned over and kissed
me, her warm, moist tongue slipping into my mouth. I looked
around the cabin at the stoned faces of the *damas* and *chambelanes*.
It seemed they were too high to notice the kiss. Or maybe I was
too high to notice that they noticed.

The entire hall was a sea of blue and white. Children jumped on
the stage and swung their arms, attempting to dance. They tried
touching the mariachis' instruments. Families gathered around
their tables drinking soda and beer. Men lined up in front of the
open bar while the beer *padrino* sweated bullets. Old ladies placed
the tables' crystal centerpieces inside their large purses (even
though they weren't keepsakes). The cake *padrino* guarded the
four-layer masterpiece, preventing the children from using their
fingers as spoons. Adolescent boys flirted with adolescent girls.
Drunken older men stared at the teenage girls—a little too long.
Young boys and gangster wannabes tagged inside the bathroom
stalls. Young girls flocked to the restroom mirror. Couples danced
to the quick mariachi beats. The kitchen bustled as the wait staff
took plates of beef, beans, and rice to every guest. More than three
hundred people filled the hall, and more were sure to arrive.

Yvette's plate was left untouched next to mine. When I looked for her, she was sitting with her father on the other side of the hall. By now his stony face wore the warmth of a father's smile. Yvette looked happy with her dad by her side. Her father was here on this important day, and it meant the world to her. They shared his plate of food while her father's hand rested on her shoulder. He said something that made Yvette laugh.

By that point I felt bad that my parents weren't present. Mom had gotten the hint for once. As Yvette stood up from her father's table and made her way across the hall, I thought that my mom would have liked to see Yvette and me together. Yvette looked the way my mom had looked when she was young: innocent. Yvette glided across the hall, the form-fitting gown hugging her thighs as she moved. She looked like the kind of girl that I would be proud to introduce to my mother. She sat down next to me, placed her hand on my shoulder, and told me that I looked great in my tuxedo. It was something a parent or sibling would say.

"When was he released?" I asked Yvette.

"Last year," she said and took a sip of champagne. "Funny, he says he wants to work things out with Mom."

"Maybe someone should tell him she's a lesbian," I said.

"Maybe he thinks he can change her. She wasn't always a lesbian, you know."

I gulped my champagne and looked around for Ingrid. She was nowhere to be seen.

"Not going to happen," Yvette said. "I love Dad. I love Mom. But I don't love them together."

Chuy's eyes were bloodshot. There was a thick glaze over them as if he had just awoken and still hadn't realized that the dream was over. The stench of weed surrounded his body. He took me aside and asked me to go over the waltz with him. I assured him he would do fine. But under the influence of those natural herbs, he said he lost his memory and couldn't remember the waltz. Large beads of perspiration clung to his face.

We found an empty area in the parking lot near the limousine. Chuy placed his arms around me. I pushed him away.

"What the fuck are you doing?" I said. "Don't touch me! There are people around. People, man, people."

Chuy danced with his imaginary Juanita. He had truly forgotten the waltz. I reminded him of a few basic steps and eventually it all came back. As usual, his upper body remained frozen stiff, while his knees and legs moved quickly beneath him, his face staring straight ahead as he strained to avoid looking at his feet. He smiled with relief, we bumped fists, and he hurried into the hall.

As I made my way toward the hall, the limousine window rolled down with a mechanical hum. Ingrid was sitting inside with a bottle of champagne, smiling at me.

As Ingrid and I hurried to put our clothes back on, I looked back and saw the two empty bottles of champagne rolling on the floor. This on top of the half dozen beers she had finished at the open bar—at the expense of the beer *padrino*.

It would have been easier had she been sober. It would have been easier had she not been upset by her ex-husband's presence. My words would have sounded all the better had we not just had sex. But as I attempted to fix my tie like my mother had that morning, I said, "Ingrid, I don't think we should see each other again."

She stopped dressing. "You just fucked me. Did you feel this way before we fucked or just after?"

"Before and after," I said.

"And during?"

"I was trying to enjoy the moment."

I tried to explain, but no matter how I went about trying to end the affair, Ingrid was not taking it well. I tried making her see the big picture. I discussed how I felt guilty about cheating on my girlfriend, telling Ingrid that I wouldn't want the same from Catalina. I assured her that I didn't regret one moment of our relationship and would always remember and treasure her. She nodded slightly. I felt that things were understood.

Then I made the mistake of telling her that Yvette kissed me. Worse, I confessed that I was really seventeen. She turned pale and vomited onto the plush black leather seats and the soft gray carpeting. I stepped outside the limo and hurried toward the hall.

Catalina and I moved to an empty corner of the hall. When no one was standing within earshot, I told her my version of the truth, explaining that Ingrid was drunk, that she vomited, and that after asking me to sleep with her, I turned her down. "I think things might get crazy," I said matter-of-factly. "Maybe you should go home, wait for me there. As soon as the waltz is over, I'm leaving."

Catalina looked at my eyes. She looked at my tuxedo, examining my tie closely. "We'll leave together," she said.

The *padrino de flan* strolled from table to table handing out the dessert. He was a dwarfish, heavyset man who wore a fedora not for fashion's sake but to cover his premature bald spot. He proudly explained how the flan was made from ingredients imported from Mexico. "It is perfect," he said. "I hope you enjoy." When he reached the main table, he handed Yvette and me the two largest pieces. "Just for you," he said. Before excusing himself to another table, he commented on my resemblance to a famous actor.

Minutes after the serving of the flan, the guests poured champagne into their glasses. The moment of the toast approached. Esteban explained to the *damas* and *chambelanes* that after the toast, we would meet outside the hall for the waltz. Esteban then made his way to the back of the hall to speak with Ingrid.

Yvette said, "Mom's a little drunk."

"Is she?" I asked.

The *padrino de honor* spoke first. He stood near the DJ's booth and spoke into a microphone. He had known Yvette since she was a little girl, he said. He had seen her grow up. He wished her luck in school and in life.

Yvette's father grabbed the microphone. He had a thick smoker's voice and sounded like an old gangster. There was an edge to his voice. He spoke about how he had not set the best of examples for his daughter. He said she grew up without him. He said that despite all of her hardships, she had stayed on the right path and had grown into an intelligent and respectful human being. No matter how many candles are on her cake, whether fifteen or twenty-five, she would always be his little girl, and he would always love her. The guests applauded.

Ingrid pulled her short dress up, making it even shorter; she probably thought it was too long and would get in the way as she walked. She stumbled as she took a step but quickly played it off, placing a hand on a guest's shoulder. She greeted the six other guests at that table. Esteban approached and helped her take the final steps toward the DJ table. He handed her the microphone, and she kissed his cheek. The microphone caught the kiss, and there was a loud pop on the speakers.

Ingrid slurred her first words on the microphone head, but it sounded as if she said, "This berry special day for my dot her." She looked at Yvette. "You are grown up. You are a woman. As you move from child to woman, be ready. The best advice: Be careful. Men are dogs. Men are liars. Look at that liar over there." She pointed to Yvette's father.

The guests fidgeted in their seats and whispered to one another in soft hums.

"He lied to me," Ingrid said. "He abandoned us. I worked two jobs, went to school to learn English, and took care of you, baby. Your daddy was a *pinche* liar."

Esteban hurried to the DJ table. He tried grabbing the microphone from Ingrid's hands. She pushed him back.

"And this little boy," she said, looking at me. I hoped no one realized she was looking at me. I sunk down in my seat. She raised her finger and pointed right at me. "This little man . . . is a liar, too. All liars. He told me he was eighteen."

The guests gasped. Yvette looked at me.

"You're right," I whispered to Yvette, "Your mom's a little drunk."

Ingrid stared at her daughter. "He told me there was nothing between you and him. So I slept with him."

The young guests exploded into laughter. Chuy yelled out, "This is some Jerry Springer shit!" The adults began speaking freely to one another and the quiet hum soon became a loud roar.

"I thought he was eighteen," Ingrid continued. "I didn't know you liked him, Yvette. I didn't know you kissed him. I love you and would never try to take your man. He lied to me. Men are dogs. *Perros.*"

Esteban and the DJ managed to pull the microphone away. Ingrid screamed at them. Esteban waved to the mariachi band and they started playing. Despite the beautiful strumming of the guitars and the soothing trumpets, no one could ignore the scene before them.

My girlfriend made her way toward the front of the hall. I readied myself. She would either slap me or throw her drink, I thought. She stared at me, her large brown eyes looking at me as if

I were the worst of the worst. Shame and guilt flushed my face. She did not stop in front of my table, as if I were not worth the time. Instead, she walked toward Ingrid.

Catalina grabbed Ingrid's hair and slapped her hard across the face. Ingrid, too drunk to feel the slap, did not flinch. Ingrid returned the slap with equal force. Both women pulled at each other's dresses and hair, slapping each other and screaming bilingual curses.

"You bitch."

"*Cabrona.*"

"How could you?"

"He said he was eighteen."

"He's *my* boyfriend, you fucking *puta.*"

"*Desgraciada.*"

"I thought you were a lesbian."

"*Ay*, my hair!"

Esteban and the DJ tried to separate Ingrid and Catalina. Yvette ran toward them with a look of shock on her face, terrified that someone was hurting her beloved mother. She held her gown up above her knees as she made her way over.

Catalina and Ingrid continued to pull at each other's hair.

"Let go of my hair."

"You let go."

"No you."

"You first."

Yvette attempted to pull her mother away from Catalina. She managed to pull Catalina's fingers out of her mother's hair. The fighting stopped. Ingrid began thanking her daughter for the help.

But before she finished, Yvette slapped her mother hard across the face, causing the crowd to flinch. Then Catalina struck Yvette in the face. A young boy with a video camera approached the three women. The boy's mother yelled at him to stop filming and get out of the way, but the boy's father kept repeating "Funny Home Videos, Funny Home Videos."

Each woman took her turn at the other two. The beer *padrino* started taking bets from the other *padrinos*. A woman took my picture to capture my expression. I tossed an empty champagne glass at her. The young boy aimed his video camera at me. I could tell that he was zooming in toward my face. I smiled nervously at the camera; I didn't know what else to do.

The DJ and Esteban moved back, fearful of getting caught in the middle of this triangle of violence. Esteban was accidentally hit and he cried out, "Which one of you bitches hit me?" He flared his arms in the direction of the women, but when the DJ pulled him back, Esteban let out an effeminate scream. Guests yelled for the security guards. The four male guards stood around arguing with one another about what to do. The guests yelled at them to do something, and one of the guards barked back: "We're not supposed to restrain women. We can get sued." The guards just stood by and observed. Apparently, they didn't have a female security guard on staff.

The three women crashed onto a nearby table. Plates of flan slid off the table and the white creamy custard spread itself all over their clothes, skin, and hair. A river of champagne followed to wash it down.

Yvette's boyfriend, Carlos, made his way toward me with a hateful look of intent burning in his eyes. He crouched and hid himself in the crowd as he made his way around the hall. As he neared my table, he ran toward me. Chuy immediately jumped from his seat and tackled Carlos to the floor.

The DJ, Esteban, and finally, the security guards, lunged at the women and fought to pull them apart. Catalina had entangled her fingers in Ingrid's hair again and refused to let go. Yvette bit her mother's arm. Security finally managed to pull the women apart and escort them to opposite sides of the hall while another security guard helped pull Carlos off of Chuy. But Chuy punched Carlos as the security guard pulled him off.

I didn't know it at the time, but Yvette's father was making his way toward me. He slipped around the crowd, keeping his back to the wall as he traveled around the tables and guests. He approached me from behind, holding a crystal table centerpiece in his hand. The DJ got onto the microphone and alerted the security guards. Right before the vase could make contact with my head, the security guards grabbed him and dropped him to the floor. Broken pieces of crystal landed near my foot.

In the end, it was easier to stop the two guys than it had been to stop the three women.

I wanted to run out of there, get into the limo, and tell the driver to get me the hell out of that place. I looked around the hall, hoping that I would spot my father or mother's face. Guests took my picture. I used to think I was the star of the quinceañera. I loved the attention before. But now I could have done without it. I wanted them to leave me alone. I wanted them to stop snapping their stupid pictures. Didn't they have any decency? Any respect? But then I realized that they didn't care about me. I wasn't their relative. I wasn't their friend. I was just the nameless *chambelán* jackass who had ruined Yvette's quinceañera. They took my picture, wanting my embarrassment to be forever captured on film.

My heart pounded against my chest. I gripped the tablecloth, using it to nervously dry the sweat from my hands. I needed to do something to keep me from running out of the hall. I needed to do something besides just sit there like an idiot. With one trembling hand holding the fork, the other shaking hand holding down the plate, I managed to cut into the flan to taste it. The *padrino de flan* was right; it was pretty good.

***

After the dust and flan had settled, it was decided that the waltz would take place as planned. Ingrid had been escorted out of the hall by Security and friends, along with Yvette's boyfriend and father. Catalina did not need a security escort; she was happy to leave on her own.

The *damas* and *chambelanes* met outside the hall and got into formation for the waltz. Yvette stepped out, looking terrible. I couldn't tell if it was sweat, spit, or champagne glistening on her skin. Her hundred-dollar hairdo now looked like a hair product "before" picture. Her mascara was smeared across her lips and cheeks. Her gown was drenched in champagne and looked as though it'd been pulled from the washer before it could dry. Someone had stepped on the edge of her gown and torn it. One of her shoulder straps had been ripped off.

She stared down at her dress, burst into tears, and ran to the restroom. Esteban and the *damas* followed after her. After a few minutes, a dama returned and asked the *chambelanes* to change into our street clothes. We all had brought our regular clothes to change into once the DJ music started.

When Yvette returned in her tight jeans and blouse, Esteban exchanged words with the DJ and soon enough, Chayanne's waltz began to play. The DJ mixed in a loud thumping bass, managing to give the waltz an urban techno feel. The couples made their way inside and Chuy turned to look at me. We thanked each other with a smile and a nod. He and Juanita marched into the hall.

Yvette wrapped her arm around mine. Small pieces of flan were still stuck on her face and neck. I grabbed a piece from her face and put it into my mouth.

"Pretty good," I said.

She grabbed a piece from her hair and tasted it. "Yeah, you're right."

We smiled for a moment. Her smile slowly turned to a frown. I gave her a friendly peck on the forehead.

"You'll do great," I said. "You're beautiful."

She squeezed my arm to thank me.

"I'm sorry . . . about all this," I said. "I ruined your quince. This is my fault. I didn't mean for any of this to happen."

"It wasn't just your fault," Yvette said.

"A lot of it was."

"There's always my sweet sixteen."

Esteban said, "Okay, you're on. Break a *pierna*."

Yvette and I entered the hall. The *damas* and *chambelanes* had formed a full circle on the dance floor. As Yvette and I entered the circle, we spoke to each other from the corners of our mouths.

"You know it was just a kiss, right?" she whispered.

"You gave me your tongue," I whispered.

"I was high. Besides, I don't think Banderas is attractive."

"Older women seem to think so."

We stopped in the center of the circle, each *chambelán* turned toward his partner, and we began to waltz. Droplets of champagne dripped from Yvette's hair as Chayanne's "*Tiempo de Vals*" thumped its hypnotic base.

Yvette forgot all about the crowd, the flan, and the way we looked in our street clothes. It was all about the waltz. I thought about what Ingrid had said about the waltz, that the waltz was the way of life, that the waltz was how the quinceañera announces to the world she is ready to fuck. I agreed with the first part.

Some of the couples danced better than others. Some of the couples didn't make good dance partners, while others did. Some of the couples stumbled through a few of the steps. Our waltz wasn't perfect, but neither was life. Through the waltz, the quinceañera announces something to the world. That night, we both were. We announced that we had made mistakes, but despite those mistakes we were going to rise up, assume our responsibilities, and carry out what we set out to do. The waltz was Yvette's way of forgiving her mother, forgiving me, and saying that the party must go on.

Chuy smiled at us, as did the other *chambelanes* and *damas*. We danced gracefully, our feet moving to the waltz. The cameras flashed and Yvette smiled big again.

My cousin Ilene answered the door and let me into her house. "I changed my mind," I said. "It would be an honor to be a godfather at your quinceañera." She threw her arms around me and thanked me. Her father brought out some coffee. We sat together at the kitchen table. Rodrigo said he knew why I was reluctant. He had heard all about Yvette's quinceañera years earlier.

"Did you see the pictures?" I asked.

"I think Yvette had the pictures confiscated before anyone saw them," he said.

We laughed about it.

My uncle Rodrigo said Ilene's quince would be expensive. He had offered to buy her a car instead, but she preferred the party. She can always have a car, he said. She will have many cars in her life. But she will only have one quinceañera.

I thought about how Yvette's quince had been as much my coming-of-age as it was hers. It was my quinceañero. Granted, I was seventeen, but they say that boys mature later than girls. I had been so ashamed and embarrassed about that quinceañera, I never thought that I would one day look back and laugh about it.

I wondered about Ingrid. When I was young, I thought I had taken advantage of her. But looking back, maybe she had taken advantage of me. I had wanted to grow up fast, wanted to be a man so soon. If life is a waltz, then we need to enjoy each turn, savor each step, and not miss out on a single beat.

I finished my coffee and got up to leave. "So what kind of godfather do you want me to be?" I asked my cousin Ilene.

"Oh, *padrino de flan*."

Love Rehearsals

BY Angie Cruz

*J*unior Martinez was his name. He was seventeen, looked just like Robbie Rosa from Menudo, and had a washboard stomach. The first time I saw him I was at Yoyo's house. She was my best friend in eighth grade. We had recently graduated and were miserable about it. It was our last summer together before high school. She was continuing her studies on a Catholic school track and I was attending New York City's La Guardia High School of Music and Visual Arts and the Performing Arts, where kids dyed their hair blue, carried skateboards, and smoked cigarettes in front of the school's entrance.

This fact did not impress my mother. If it wasn't for the fact that art school was free, I too would've been forced to go to Catholic school. In this case being broke had its benefits.

On the day I went for my audition at La Guardia, my mother held on to me by my wrist. She gasped at the sight of three teenage boys loitering in front of the school wearing combat boots in 90-degree weather, their eyes lined in black eyeliner. I shrunk in fear that she would embarrass me in front of them even before I knew if I would get in or not. Needless to say she paced outside the classroom for hours while I went through the arts exams. Afterward she lectured me about how important it was that I kept my own personality and that just because someone jumped off a bridge did not mean that I should jump off a bridge too. She said that *los blanquitos* could act their way, but we were decent people and there was no reason I needed to dye my hair strange colors like them.

Yoyo and I weren't looking forward to high school. We knew that things would never be as they once were between us. It would become difficult for us to see each other on a daily basis. We would inevitably make new friends. So we promised ourselves that we would make the summer of '86 the best ever. We juiced every privilege junior high graduates have, and begged our proud mothers to buy us new dresses and heels. We could finally wear makeup without hiding it from them. We were also issued our first official work permits.

That summer, Yoyo and I worked as youth counselors in the city's Summer Youth Program, and on weekends I worked at the Dunkin' Donuts in Penn Station. It was my way of helping my mother pay for the graduation expenses. After work we usually hung at Yoyo's house clipping *Seventeen* magazines, looking at dresses and hairdos.

That's when her cousin Cynthia barged into the room, upset that she didn't have enough girls to perform in her quinceañera. She needed seven guys and seven girls. She had already recruited all her family, including that hot cousin Junior of hers who just arrived off the boat. But she was short one girl. "What am I going to do?" she cried and flopped on the bed over our carefully laid out pages of the perfect eyebrow shape, the best waistline for your body type, and what to do when you're not ready to go to second base.

"Angie can do it," Yoyo popped up. Her big and brown eyes opened wide as she nudged my arm to say yes.

"Really?" Cynthia clapped in delight and ran off.

Yoyo and I jumped up and down thinking about what better

way to spend our last summer together than to be part of a quinceañera party. Neither of us had been to one before. This would be our first.

Cynthia walked back into the bedroom. She lowered the music.

"Hey, Angie, I'll need one hundred and fifty dollars next month for the dress, shoes, and umbrellas. You can do that, right?"

"Well . . . I have to ask my mother. I mean—"

"And we'll be rehearsing on 137th Street every Saturday. So don't make plans. Everyone has to be there. The choreographer insists on it."

"Choreographer?"

My heart pounded against my chest and a lump rose up inside my throat. My mother would never allow it.

"So you'll be there, right?"

I nodded my head yes. I meant no. I worked on Saturdays. My mother had bought me the dress that I wanted for graduation and I promised to pay her back slowly. That was the deal. It was a white Jessica McClintock from the Misses section in Macy's. I tried on thirty dresses until I found the one that made all the baby fat scrunch up and out in all the right places. As long as I didn't smile (I wore braces) I looked pretty good. I was also saving up for all the supplies for art school I'd need. I had aced my entry exams, after all.

I should've waited to discuss it with my mother. But it was too late; Yoyo was dancing around the room and I was dancing with her as if it were a done deal. Pimpinela was blasting in the background. We held hairbrushes up to our lips like microphones and sung along to the Argentine brother-and-sister duo. We promised ourselves that we would always love Pimpinela, and we would always be best friends.

So I did what every daughter must do to get her way with her mother. I became the daughter that she always wanted.

"I gotta go." I rushed home before my mother arrived. My brother didn't even look up from the television to greet me. I started dinner. I took out the steaks from the freezer, put the water to boil for the rice. I swept the floors and pushed my brother into the bedroom while I set the table. I put on the Lionel Richie cassette, because Lionel had a way with my mother. He always put her in a good mood. I sprayed the house with Pledge. Then I sat down with the encyclopedia opened in my lap. She would be suspicious to see me reading in the summer about anteaters, but there was nothing else to read except for *Wifey* by Judy Blume, which I read only in secret.

My mother worked during the day, seven days a week, and went to the university on most nights. She was often tired. So she was always happy to see dinner on the table.

"No, whatever it is, no." That was what she said when we sat down to eat.

"But I haven't asked you for anything."

"You will and the answer is no."

"Okay, what if I tell you that this will be the last thing I will ever ask you for in my life and I promise to do everything you want, as long as you let me be part of this quinceañera this summer. It doesn't even get in the way of school because I don't start till September and the party is like at the end of August."

"How much is it?"

"Free, I think. This girl's father owns a bodega and she's paying for everything."

"Really? What bodega?"

"Somewhere on 137th."

"He's Boricua?"

We're Dominican.

"Dominican. Maybe Cuban." Boricua would've been the wrong answer and Cuban the best answer but too far from the truth. Cynthia's family were actually Dominicans who lived in Puerto Rico before they moved to the United States.

"Aren't you too young for this? You're only . . ."

Then something strange happened. She looked at me and then at my boobs, which were huge. They had popped out like two small grapefruits overnight. "You're fourteen. My goodness, you're fourteen."

She stared at my face and I worried that yet another zit was surfacing as we spoke. Zits the size of grape tomatoes erupted in random parts of my face. It was summer, the fight between my blown-out curls and the humidity had begun. My bangs, dry and brittle from too many hair relaxers, shrunk up like a small bush above my eyebrows. My teeth bucked out, pushing up against my upper lip. My jaw had tensed up from the pain of having my braces tightened every other week. I was going through an ugly phase and her looking at me for so long made me fear that something else was going terribly wrong with my looks.

"Fine. If they're paying for it, then fine. But only if it doesn't interfere with your job. I need your help around here."

Rehearsal wasn't chaperoned. It was in a high school gym where one of Cynthia's uncles worked. I had never been in a high school gym before. It was enormous and felt dangerous. There were dark spaces behind the bleachers and the smell of boy funk. Quickly I understood that Cynthia's family had connections. Her father's bodega was like the heart of the block and everything near it was open to her use. You would think she would have had more friends.

We all were in awe of the choreographer, Jewel, who was nineteen and claimed to have been in the "Thriller" music video (she was one of the twitching corpses without arms). She also said she toured with Lisa Lisa & Cult Jam on occasion. Jewel wore one black lace glove. She claimed it was the same one that Lisa Lisa had worn on her album cover. Jewel swung around a stopwatch, adjusted her leg warmers often, and carried a wooden stick that she pounded on the floor to keep time. She made us stand in a straight line, one behind the other. Junior had taken the place behind me. That made it difficult to pay attention to anything Jewel was saying.

I had seen him once at Yoyo's house from afar. But when I felt his breath on my neck my heart raced and my hands got clammy.

"Back straight, shoulders back, chin up, and smile." She said as she walked up and down the line tapping our butts in, rearranging our order. I refused to smile. I got used to covering my mouth when I had to. I didn't like looking at a mouth full of braces and didn't expect anyone to have to look at mine.

In front of me was Yoyo's boyfriend, Elvis. Yoyo had been dating Elvis since she was thirteen. I hadn't dated anyone yet. My mother had me on a tight leash. Not like Yoyo's mom. Her mother preferred knowing what her daughters were up to. Elvis introduced himself to Yoyo's mother and asked for permission to see her. Very old school, very proper. She even allowed Elvis to enter Yoyo's bedroom and close the door. But even with all that freedom, Yoyo wouldn't allow Elvis to go to third base. Above the waist was the rule.

Unlike Elvis who liked to make silly comments and tease Jewel, Junior was quiet, intolerably quiet. I couldn't tell if he was bored, shy, or arrogant.

"He doesn't really speak English," Yoyo whispered to me as if she could read my mind.

"*Hola,*" I said to Junior. I turned around and waved at him. It was stupid but out of my control. When I got nervous words flew out of my mouth.

"*Hola,* " Junior replied. I looked back, with my hand over my mouth and giggled. Dimples emerged on his cheeks and his hand went down to rub his stomach, then up into one of those deep long yawns that boys do when they want to show off their abs.

That's when I realized that all the lying to my mother was worth it. All the extra hours I would have to work at Dunkin' Donuts to raise 150 dollars was going to be worth it. All the denying myself Suzy Qs, Ding Dongs, Oreos, and Frosties just so I could look thin and pretty on the day of the quinceañera so Junior Martinez would smile at me, was worth it.

In some ways, Cynthia's quince was the closest thing to a quince that I would ever have. There were going to be cameras, a live band, and we were going to be performing for an audience. All that decadence, when we were all scrounging for pennies. It was a big deal for everyone. My mother could never afford a party like Cynthia's; I wouldn't even think of asking her for one. Years later I learned that Cynthia's family couldn't afford it either. Her father took a loan against his bodega just to please his daughter.

Cynthia was her father's princess. And perhaps in his mind that made him the neighborhood king. He owned a bodega. People depended on him. He gave people food on credit. And unlike many of us, Cynthia had a father that stuck around. None of us had any idea what her everyday thoughts and desires were because she kept to herself. The few exchanges I had with her were when she asked for money or said what we should or shouldn't do at her quince. "*Mi quince* " was how she referred to it. "*Mi fiesta.* " And we were her backup, doing everything we could to make *her* look better. Besides, Yoyo and I were little girls in Cynthia's eyes. In a few weeks she would be a young woman. And she did look a lot different than us. Even with those few months she had on us in age, she looked more experienced, more mature.

Cynthia was pretty in a Barbie-esque way. She had a long black mane that she proudly called *"pelo bueno."* Not like my hair, that was *"pelo malo"* or *"pelo vivo."* She parted her thick and shiny mane to one side and a big swooping wave framed her face. It bounced and moved around her in the most perfect way. And her skin was pale with white fuzz around her cheeks, as if she had never seen the sun before in her life. She had long black eyelashes that she constantly extended with Maybelline mascara, so she always looked bright-eyed.

Jewel, the choreographer, took her job seriously. She gave us a diet: no sugar, no bread, no rice. Then a uniform: We should wear black to rehearsals so we would look like a team. I hid the Dunkin' Donuts uniform deep down in my bag. Saturday rehearsals were all day and I told my manager at work I had family business every Saturday until the quince. I took the risk that my mother wouldn't go to check in on me. She was the type to do that. She always assumed that I was lying or doing something behind her back. That's what America did to good girls like her daughter. And she was right, I was lying to her.

When I saw her in the evenings, and she greeted me as if I were the good daughter arriving home from a long day's work, I felt terrible betraying her. But at the time I thought she would never understand. Besides it wasn't like I was doing anything bad. Just learning some cool dance moves; gyrating like Madonna, moonwalking like Michael Jackson.

The choreographer's plan was that Cynthia would strut into the room to a condensed version of Run-D.M.C.'s "Walk This Way," when it would suddenly mix into "In Your Eyes" by Peter Gabriel. This was when us kids in the court were supposed to go from strutting to some sort of waltz to a short step dance, with some very synchronized clapping and stomping action. Then we'd go back to bumping to some house music and then all formal again for our folkloric merengue number. It was an old-style merengue where the men bowed and the girls curtsied. All this would eventually lead into Wham's "Careless Whisper."

That's when we would all come together. Our bodies huddled, building a large human cocoon. Cynthia would be hiding in our circle crouched low and we would then open up and she would lift herself slowly like a flower blooming and look around to find her father, who would then walk her around the room to look at all the guys waiting to dance with her while he watched his little girl leave him and become a woman.

"It'll be amazing," Jewel assured us and asked us to trust her. "I've done this for a very long time."

Back then Jewel's nineteen years of life felt like a very long time. She flung her fake ponytail left and right while she waved her hands around like a maestro. She always pointed at our cheeks and poked her finger under our chins, reminding us that when we danced we should be smiling and lifting our chins; we should be proud of who we are.

The choreographer only rehearsed with us in the mornings because she said she had better things to do than spend her afternoons with us. She was getting paid by the hour and once she completed her hours she was off. For lunch Cynthia's father made us sandwiches from his bodega and let us have any soda of our choosing.

Yoyo and I sipped Coca-Colas and ate ham-and-cheese subs for lunch, M&Ms for dessert. Unlike Cynthia, we didn't think about Jewel's diet. Cynthia was thirty pounds overweight and was on a mission to lose it by the day of the event.

"Impossible," Yoyo said. "She doesn't eat all day and then she eats like a gallon of ice cream at night because she's starving. She's lost her mind. Did you know she's getting diamonds sewn onto her dress?"

"Real ones?"

"I don't know. She spends all her time looking at herself in the mirror afraid of zits and things."

Even though Yoyo felt free to be critical of Cynthia, I never dared say anything bad about her because family is thicker than blood and if Yoyo had to choose between Cynthia and me, she would choose her. There was something heartbreaking about that fact, but it was a reality I knew from my own family.

As the summer days went by, Yoyo started distancing herself, preparing us both for the fact that we'd no longer be walking to school together. We couldn't even ride the train together. She would be going to school up in the Bronx while I'd be heading to midtown Manhattan. After six years of being inseparable, we would be facing the world alone.

So when Junior finally talked to me at the next rehearsal I was ready. Yoyo was making out with Elvis while I stood under the EXIT sign like an idiot waiting for her to finish.

"You need a ride?" Junior asked me.

"Home?"

"Yeah." I tried not to look excited in front of Yoyo because then she would know I was into her cousin and make a fuss. She couldn't keep a secret. And I didn't want him to know how much I liked him. He was seventeen, but not hard-looking or immature like the seventeen-year-olds from around the way who teased me in front of their friends and then tried to kiss me in the back alleys when their friends weren't around. Junior was different; he was a newbie. He had just arrived from the D.R. and was about to finish his GED and attend a community college.

"Go with him, *loca*," Yoyo said, looking up from lip-locking with Elvis.

"Don't you want to come?" I said.

"Elvis and I are gonna hang for a little. Go," Yoyo insisted and then asked, "Don't you think he's cute?"

"How did you know?"

Yoyo took my hand like the wiser soul. My heart was thumping so hard my chest hurt. I turned beet red.

"Everyone knows you two are crazy for each other."

I looked around at everyone, who were busy doing their own thing. Had they been talking about us? How could I have been so obvious?

I took my knapsack and followed Junior. He took me to his car. It was a dented Toyota Corolla begging for a paint job.

"Excuse the car. I've been working on it, but it takes awhile."

"No, it's okay."

That was my mother's worst fear: me in a car with a raging adolescent boy who probably didn't have a driver's license. And her daughter's hormones were also raging. I was hot all around my hips, holding my lips together to look as pretty as possible.

He opened the door for me and then got in the car. He bent over to get something from around my feet. He was feeling around, caressing my legs, and then found his sunglasses.

"Where do we go?" Junior asked.

I told him to leave me in front of a small park near Columbia Presbyterian Hospital. It was a few blocks away from my apartment on a street my mother wouldn't walk on. When we arrived I stood outside for a moment while he sat in his car looking up at me. We stared at each other for a while waiting for something to happen, for something to be said. I held my knapsack tight in front of my chest as a barrier.

"I can pick you up at work tomorrow. Yoyo told me you work in Penn Station."

"I finish at three."

"You're cute," he said and again his dimples emerged and I tried to hold myself together.

Junior picked me up at work the next day as promised, holding Popsicles for us. We ate them while he zipped uptown like a maniac driver and we talked about the Dominican Republic. We shared how much we liked it down there and agreed that if it wasn't for our families, who insisted on living in New York, we would be just as happy if not happier in the Dominican Republic.

"You're a good listener," he told me when he dropped me off and asked me when I had to work again so he could pick me up. I saw Junior four times that week. He never made a pass at me. He never did much more than feed me Popsicles or *frío frío* drinks and drop me off.

Then Saturday came and we were at rehearsal. But it wasn't the same as it had been before we talked, before I'd been inside his car. Yoyo knew that I was seeing Junior. I told her he was just being nice; she said it was more. Junior had been asking Yoyo questions about me.

"Like what?" I asked her.

"You know, questions. Like what kind of family you come from, and if you like to study. And if you're a good person."

"What do you tell him?"

"That you're the best, that's why you're my friend."

Cynthia was late to rehearsal. She was sick to her stomach. Literally. Yoyo said that Cynthia had taken some pills to lose her appetite and that there had been side effects. She told Yoyo to tell us that we should still rehearse without her. The choreographer decided to split. She left us the tapes and told us that we should go through the songs and get to know each other. The best dancing happens when people are comfortable. She walked out with her usual end note, "Don't forget to smile." Her words were directed right at me because I never ever smiled. Unless someone made me laugh or when I looked over at Junior, of course.

"Hey, cutie," was what he said when he found me sitting alone on the bleachers listening to my Walkman. I was waiting for Yoyo to return from the bathroom.

"*Hola.*" It wasn't easy for me to get past *hola.*

"Do you need a ride home today?"

"You don't have to always take me home."

"I want to," he said and took my hand into his and looked at me with his soulful eyes. I puckered my lips and pointed them toward him when Dionne Warwick's "That's What Friends Are For" came on and Yoyo screamed, "That's our song!"

She pulled me to the dance floor and pretended to slow dance with me, until Elvis pulled us apart and said to Yoyo, "Stop looking like a lesbo." I don't think Junior understood any of it.

We danced through our quinceañera routines as usual, but all the songs took on new meaning for Junior and me. When we slow-danced to "Careless Whisper" my body and his were one.

That afternoon, it started to rain and we were caught under the ledge of the building. His car was parked a few blocks away. He held my hand while we waited and our bodies inched closer and closer to each other until he turned and planted a kiss on me—a kiss that lifted me two inches off the ground. I thought I was floating until I realized that he was lifting me. We were kissing so hard, his tongue deep in my throat. It was my first real grown-up kiss. The rain pounded on the sidewalk, just inches away from us, his body protecting me from the rain, his back soaked.

I returned home in bliss, soaked from the rain, smiling from ear to ear. My mother was home.

"How was work?"

"Good."

"You never have that donut smell when you get home like you used to. Strange, no?"

"Yeah, must be the sweat. It's like taking a shower. And then I got caught in the rain."

"I went by your work. I spoke to your manager. You haven't worked on Saturdays for the past three weeks. He said you have some family commitments."

"Look, Mami—"

"Don't 'look, Mami' me; do you know the sacrifices I make for you children? Every day I am out there trying to keep food on the table and you're out there doing Lord knows what."

"Mami, I was going to tell you. I just didn't want you to worry. I mean I had to rehearse for the quinceañera on Saturdays, but because we needed the money I didn't—"

"The quinceañera? You lied to your mother for a quinceañera?"

"I'm so sorry, Mami. I promise—"

"No sorry, no nothing. You won't be doing the quinceañera. Call that girl and tell her that you quit the quinceañera. And from now on, I will be picking you up from work. You obviously can't be trusted to be on your own. I thought I could trust you but—"

"I'm sorry, Mami, please . . ."

She picked up the phone and held it in front of me, "Call her. Now."

I called Yoyo. I didn't even have Cynthia's number. I told her I couldn't do the quinceañera. I thought at that moment that my life was over. I would never see Junior again.

That night, Cynthia's mother called my mother and said that it was unfair to Cynthia if I didn't participate because the choreographed dance required seven girls and it was too late to find someone new to replace me. She said my mother was welcome to attend rehearsals if that made her feel more comfortable and would she please reconsider. Then she mentioned that the fabric for the dresses had already been ordered and that we would lose the fifty-dollar deposit I had given her.

"Deposit for the dresses?"

My mother clenched her hand into a fist and pressed it on her hip while she talked on the phone. She was trying to stay calm. Worse than my betrayal of ditching work was that I spent money behind her back. We were supposed to be a team, a family. We were supposed to help each other out first before we spent our money elsewhere.

My mother hung up. She was quiet for a long time; then in the calmest voice I ever heard from her, "You have betrayed me. I didn't think you would ever be this kind of daughter. You can do the quinceañera because it's obvious that you will do as you please. I hope you can live with the fact that I'm heartbroken and that I can no longer trust you."

I told Yoyo to tell Junior not to come by work anymore. He would have to wait until the next rehearsal to see me. I did everything I could to make my mother look at me the way she had before. I didn't talk on the phone, I kept the apartment neat, I cooked dinner, I helped my brother with summer school homework. I tried everything to please her, but our relationship had changed.

Cynthia showed up with the swatches for our dresses at the next rehearsal. Butter yellow. The kind of yellow that was unflattering to every complexion and shape in the room. The kind of yellow that only anorexic Hollywood blondes could get away with.

"Don't you love it?" she said and then showed us her dress color. A ripe peach color. One that accented her dark hair and the pink in her cheeks. The men would wear white tuxes with complementary yellow bowties and vests.

"Yellow?" My disapproval spoke for everyone.

The fabric had been bought. The design picked out. It was a light chiffon dress with ruffles around the neck and on the hem and a taffeta sash around the waist. We would wear flowers on our heads, possibly a headband. The shoes would be dyed to match the dress and umbrellas and, of course, we would have to pay for those, too.

"You'll look pretty in yellow," Junior said to comfort me.

"Thanks."

When I looked into Junior's eyes I didn't regret any of it. That was dangerous. It made me wonder what I was capable of doing to keep looking into his eyes. My mother started picking me up at the end of rehearsal, so Junior and I spent lunch break behind the bleachers kissing. It was the only time we had for ourselves. We kissed for thirty minutes nonstop. We skipped lunch. Who needed food? We were in love.

When we danced, we did everything in our power to caress something new on the other's body; sometimes it was a pinkie, another time it was brushing past him close enough to graze his crotch. My mother was right to be worried.

After six weeks of rehearsals our group of fourteen became a family. We freely teased Cynthia behind her back, mimicking her demands as to what we should do or not do at her party. By week six, everyone had coupled off with someone in the group or with someone from the block.

Everyone except for Cynthia, who had no time for love that summer. She was losing weight. Not the thirty pounds she hoped for, but the pounds were coming off and the thinner she became, the more neurotic she got. But none of us cared, especially Junior and I, who hardly exchanged words. We just kissed and looked into each other's eyes. That was enough to communicate how important we had become to each other.

The big day was fast approaching so we met more often to help decorate the gym with paper streamers in peach and yellow. Even my mother became more excited when she went with me for the dress fitting.

"Maybe we can get those braces off before the party. It will be better for the pictures."

It was premature and my dentist was opposed to it, but after two and a half years of braces I had had enough. The braces came off and I felt pretty, but naked. As if suddenly because my teeth were exposed, my feelings were too.

When I arrived at Cynthia's apartment on the big day in my yellow dress, my face said it all.

"What's wrong?" Yoyo asked as soon as she saw me.

"It's awful," I said, looking down at the abundance of yellow on my body.

"Your teeth." Yoyo went to touch them.

"Surprise."

"You look so beautiful," Yoyo said and pulled me over to a long mirror.

She stood next to me and then another one of the girls stood next to us. Together we didn't look as bad as when we were standing alone. And maybe that was the point.

"Pretty cool, huh?" Yoyo said and twirled around to show how her dress lifted in the air.

We all twirled around while Cynthia flitted about like a butterfly, nervous that the flowers in the gym were not exactly the color she wanted, that Jewel had still not arrived and she needed her to be on time to instruct the DJ. Cynthia acted like a crazy person and accused everyone of ruining the most important day in her life.

"What do you mean Jewel had to drop off her mother at the doctor? Doesn't she know that I need her?"

Cynthia's face was dewy from the August heat. We took turns patting her forehead with tissues, fetching her water, calming her down by saying things like, "We're here for you, Cynthia." And while I didn't care for Cynthia that much—she was never particularly nice to me—she suddenly looked vulnerable and needy and we all tacitly agreed to play our parts as her ladies-in-waiting. We fussed around her, adjusting the flowers in her hair and puffing up her skirt while we waited for the white stretch limo to pick us up and drive us to the Cloisters where we would take photographs. We made sure her makeup didn't bleed when she touched her face. She didn't say thank you, but it was her day and we were in our dresses and we had gone from rehearsing to the real performance.

"Angie." Junior said my name in his Spanish accent. My straightened hair was curling at the ends and a flower made of fabric was tucked behind my ear. I turned to him and smiled. He looked dorky in his white tux, which had tails. His pants were too short and the jacket too tight. But when I looked into his eyes he was my Junior.

When he saw my pearly whites his eyes widened. He nodded approvingly.

When we arrived at the gym and I saw the disco ball, the gym converted from a recess area into a supper club, crowded with round tables draped in white tablecloths and flower displays, when I saw that everyone was anxiously waiting to see what all the hype was about, I became nervous. We were supposed to put on a show for them. Junior's grip on my hand was tighter than usual.

"This is your moment. Remember to smile. You've been training for this day. Now make me look good." The choreographer seemed to have taken her pep talk lessons from watching too many sports movies. But at the moment, her words were exactly what we needed to hear. She hugged us all one by one. I think her eyes even welled up when I smiled. Finally, she had gotten a smile out of me.

The DJ announced our arrival one couple at a time. Junior squeezed my hand and pulled me close to him. "My mother's out there," I warned him, but it was difficult to hide everything that had transpired between us. We were a team, his cummerbund matched my dress. I belonged to him. And there was a part of me that wanted my mother to see that I was in love with Junior. When the DJ said our names we strutted out together and he twirled me around until I was in position. I could see no one and nothing outside of the dance floor. Bright camera lights flashed in our faces, smoke spread around our feet. The music was blasting and everyone was standing, waiting for Cynthia to appear so the show could officially begin.

Cynthia swayed into the room escorted by her father, who unleashed her. Everyone clapped when they saw her dress; she was straight out of a fairy tale. Small fake diamonds were hand-sewn onto the sheer fabric all along the hem. When she moved, the fabric glittered and it was tight around the bodice, making her waist look two times smaller. Then the skirt belled out to reveal the crinoline underneath. Her hair was swept up from her face and a cascade of curls fell on her back. Very *Gone With the Wind*.

The disco ball spotted our faces, the music and lights were in synch with our movements, and for the first time Junior and I were able to show our love for each other in public. No more hiding in bleachers. We stomped, clapped, stepped, waltzed, and sashayed without a glitch. The crowd stood up and applauded and when Prince's "Kiss" came on, everyone bombarded the dance floor.

As soon as we were done with our commitment to Cynthia, Junior pulled me outside to the hallway. We were sweating. He told me that he was in love with me. That he wanted to ask my mother if we could date. I knew my mother wouldn't allow it and I didn't want anything to ruin my night.

"Just kiss me," I said. No more metal mouth. No more bruised lips.

We went back inside and right onto the dance floor as if we had never left. When the song finished, I walked away from Junior toward my mother, who was beaming with pride. She hadn't noticed that I had walked out with Junior.

"You look so beautiful. You did so well." She spoke as if she had forgotten how betrayed she had felt after I lied to her.

When Junior and I parted ways that night he asked me to promise him that I would never forget him.

"Why would I do that?" I asked.

"You start a new school. A new life."

"Don't worry. Nothing will change between us," I promised.

A week later I started high school. Yoyo and I called each other to swap notes about how different it was. But each day our conversations became shorter. Junior kept asking me if he could pick me up at school, but I told him I would get in trouble with my mother and that it was better to wait for the weekend to see each other.

When the weekend came around and he would show up at work, it wasn't the same. There were no bleachers to kiss under or rehearsals to dance for. Outside of the quinceañera I had very little to talk to him about. When he asked me about school, I didn't know what to share with him. The new world I was witnessing was so different. It had been only a few weeks since the quinceañera but it felt like years had passed. It seemed like it had all been a dream, not connected to my reality at La Guardia High School.

Once my zits were under control and my braces had been removed I started to flirt. Boys were checking me out and I was winking back. There was one guy, Nelson, who had singled me out from day one. He was a senior who looked and dressed like a brown version of George Michael; he wore jeans that were tight around the ankles and baggy turtlenecks. He was an art major like me and one of his paintings was being exhibited in the school lobby.

"Maybe it's better we don't see each other anymore," I said to Junior. In the same way "*hola*" had slipped out of my mouth the first day of rehearsal, I broke up with Junior. He didn't drive me home that day. He walked away, leaving me standing alone at Penn Station. I went home, nervous, unsure where I was going or what I had done. Breaking up with Junior was more than breaking up with a boy, it was breaking up with a part of my life that quickly had become foreign and so different from my new world at La Guardia. I couldn't imagine someone like Junior and Nelson in the same room, let alone in my heart.

A few nights later I was home watching a telenovela with my mother when the phone rang. My mother answered; it was Cynthia's mother. She said that she wasn't sure what kind of mother she was, but she could certainly tell her what kind of daughter I was. She told her that I misled her son, Junior, to think that I was a decent girl and that after professing my love for him I threw him away for another man. She had found Junior drunk on the floor of the bathroom because I had left him. She said that after giving me the privilege of participating in the quinceañera where the Martinez family opened their hearts to me that I had proven to be ungrateful.

My mother listened while I stared up at her. She defended me by saying that I was right to refuse Junior and his bad intentions. I was too young for a boy like him or to be in a relationship with any boy for that matter. That Junior should learn to be a man and get over it.

It was the first time I heard my mother defend me in that way. For weeks I assumed she would bring it up, but she never did.

Less than a year later I was about to turn fifteen. I had survived my first year in high school. My mother was happy that my hair color was still black and that I had no tattoos or piercings to speak of.

"What do you want to do for your birthday?"

I knew we couldn't afford a party, but I remember not wanting one either. Yoyo and I became more distant after I broke up with Junior. And my new friends at La Guardia were still too new. So I asked my mother if we could celebrate at the new Hard Rock Café in midtown. I invited my two cousins, my mother, and my brother. Eating there was exotic for all of us. We had hamburgers, french fries, and chocolate cake. My mother brought fifteen candles in her purse. It was perfect.

Fifteen and the Mafia

by Constanza
Jaramillo-Cathcart

I like walking through Brooklyn because in an instant, I can be transported out of my present American life into my Third-World upbringing. You could argue that a little Third World could be found in *botánicas*, or those variety stores, which sell everything from Barbie dolls and cleaning products to Chinese underwear. But they really don't compare to Ocasiones, the little shop on Kane Street with the tulle dresses and the mini tuxedos in the window. The clothes hang there waiting for another

Latin child to graduate to that next stage in life: a baptism, first communion, or a girl's fifteenth birthday.

Every morning this summer, I have walked past the store and seen the display from the corner of my eye, but today, a white glow makes me stop. I park my baby's stroller in front of the window and look up. It's a quinceañera dress. The delicate tulle skirt hangs behind a clear plastic bag, like a wrapped meringue. I stare at it, and my mind gets lost in the cloud of fabric for a few seconds, until I realize something is bothering me. I squint and see the reflection of a telephone technician perched on a lamppost behind me. He looks as if he is climbing onto the skirt, trying to hook cables inside of it. And I can't help but think: Angel hasn't called. It's been twenty years since we went to Magdalena's quinceañera party. And he never called.

Twenty years ago I lived in Bogotá, Colombia. I was fourteen, having already spent ten years as a student of the Louis Pasteur French Lycée. This institution was mostly staffed by our underpaid national teachers or by young French people who had taken jobs in Colombia in order to go on exotic vacations and avoid the draft. And since the Colombian teachers were paid much less than the French ones, they took it easy. Take for instance, our Colombian gym teachers. In their ironed nylon track suits they'd puff on those filterless cigarettes hanging from their lips as they yelled to us, "One more lap." This in between coughs.

The French school was an old redbrick building built on the rim of Los Cerros Orientales, the Andean range that looms over Bogotá. The school had very little in the way of grass lawns, and it boasted several cement playgrounds instead. Most of the classrooms had high ceilings and large windows that looked out onto the gray playgrounds against the cobalt Andes. To our delight, large brown moths and the occasional mouse would disrupt class every so often. When I was fourteen, I was brought to an agonizing mortification by one of those moths. First period class was starting and I had just sat down, not realizing that a large brown butterfly with wings as big as our male teacher's hands had taken shelter under my desk. It quietly stuck itself to the wool threads of my skirt. When I jumped around the classroom in panic, flipping my skirt up and down and screaming, the young Frenchman snapped, *"Asseyez-vous, Mademoiselle!"* Veronique, my new friend at the time, gave me a sideways glance that said, "When are you ever going to learn?" which seemed worse than his screams.

We had become inseparable, Veronique and I. We were in the same grade and lived in the same building. I lived on the sixteenth floor with my parents and little sister, and she on the fourteenth with her mother and stepfather. According to my other friends, Veronique was a *madurada biche*, meaning "a green fruit gone spoiled." It was precisely the kind of influence I needed at the time. It takes years of reflection to fully appreciate the benefits of having a bossy and precocious friend at that stage in life.

Embarrassment. That is how I can sum up being fourteen. I don't mean to imply that at fourteen I was horribly embarrassed all the time. It was worse. It just waited there, lurked there, ready to strike at the least expected moment. Or there was that embarrassment of being embarrassed about everything, which was constant, like the permanent but almost forgettable nausea one feels at the beginning of a pregnancy.

The things that mortified me didn't bother Veronique, so as dominant as she was, I couldn't help but feel comfort at her side. It seemed I just couldn't get rid of the things that made me uncomfortable: my nose, which I couldn't take my eyes off of; my braces, which I couldn't tear off my teeth; my thick and straight hair, which teasing didn't do a thing for; and there was not knowing what to say when I needed to the most.

Veronique, on the other hand, didn't mind her braces. Her hair was curly and she could tease it to perfection with just a little touch of spray, and she always had a snappy thing to say to the boys. She was skilled in that mysterious art of saying mean things that amused the opposite sex. For some reason, they never took it personally.

She had one weak spot, however, a terrible Achilles' heel. Her last name in French closely resembled the word "penis" in Spanish. I never brought it up, but would occasionally think it when I was mad at her. At least the moth flew away and I was able to sit back down, but she, *she* would be stuck with that phallic surname until the day she got married.

At an institution where every student was considered equal our friendship was anything but. The young European staff never distinguished between the children of the faculty or maintenance staff, the granddaughters of the ex-president, the five children of the Haitian ambassadors, the sons of the bus union leaders, or the rest of us. We were all little Colombians absorbing the French language, culture, and history. They never acted surprised when we answered their questions in our flawless French accents. Nor did they realize what we would say about them in Spanish. And they endlessly complained about *ce pays*. It didn't matter what the problem was: If there was no chalk, or it rained, or the window didn't close properly, there was always "this country" to blame.

I liked it there. I liked singing the Marseillaise and the Colombian anthem on Monday mornings. I liked wearing that gray skirt and blue blazer uniform. I liked arguing in French and writing minuscule French words in notebooks that looked like music scores. I hadn't yet realized our gray Bogotá was a melancholic place. This French island in the middle of the cold Andes was my home, and I enjoyed sitting in the cement bleachers during breaks and looking up at the menacing skies.

The truth is that as students of the French school, we didn't have a clue about the country we lived in. We knew Montesquieu, Baudelaire, and Racine, but little did we know about Bolivar, Núñez, and Silva, much less what was going on in the news.

In the mideighties, the tough times had not yet begun. Even though the drug lords had already taken Justice Minister Rodrigo Lara Bonilla's life, it wasn't until 1989 when we all realized how bad things had gotten. In that year, Luis Carlos Galán, the beloved presidential candidate, was assassinated on the podium, and around then, Pablo Escobar decided to start his bombing campaign against the city.

But I don't remember the year 1986 as a scary time. You could say much of the country was still being seduced by *coca* wealth. Fabulously extravagant people were beginning to meld into society. You started seeing late model Mercedes speed through respectable neighborhoods. Stores with never-before-seen luxury goods started popping up. And you heard the stories: A man walked into a discotheque, ordered it to close, and bought Veuve Clicquot champagne for everyone. Another one replicated an entire country club in his backyard because he wasn't allowed to join. A building with a swimming pool on every floor was being erected in Medellín. And at a quinceañera party in that same city, every girl who attended walked out with a new car.

Veronique would sometimes add her two cents to the stories. "My father has a friend who is a mafioso and you just wouldn't believe his country house!" I really never knew if she was making it up, but I admired her boldness.

When I discovered there was more to life than school, thanks to Veronique, my grades started to suffer. But my street smarts started to improve. I didn't care so much about those French algebra equations. I stood up my tutors and started asking my parents for weekly manicures instead. They said yes to the manicures, but no to the nose job I also demanded. And even though my mother and I started having frequent screaming matches, I don't remember her asking too many questions or placing too many restrictions on me.

Take for example, them never knowing that Angel and I had gone together to my classmate Magdalena's sweet fifteen party. I had become unrecognizable to them in a very short time span, but they didn't seem to mind. In regards to my dwindling grades, my father declared to my mother that "maybe she just needs the experience of repeating a year." And when my mother and I would have our fights, I would immediately enlist my father in my plots to drive her crazy. Plus, he was the one who bought me my outfit for Magdalena's quinceañera.

Since the party was coming up in a few weeks, my father took me shopping. I had really wanted to get out of the house because my mom and I had had another argument, this time about my leaving plates of food and other general forms of mess sitting around the house "for her to pick up." When she dared to say that about an old molding plate of fruit salad in my room, I became irate and demanded my father take me around town because I had "nothing to wear."

I settled on an oversized white shirt with large fruit patterns stamped in black, a jumbo-size brooch in the form of a diamond heart, and a blue stretchy acrylic skirt. The shopping trip ended at the hair salon, where I got a layered haircut that I could finally tease and make stand up. My father patiently helped me get my new look together, oblivious to what this would do to my mother. When I came home she just looked at me without saying a word. There was her little girl, who a few months ago got good grades, looked innocent and, half jokingly and half affectionately, would answer her with "*Sí, señora*" or "*No, señora*," as good Colombian children are taught to reply.

At the time, I remember secretly enjoying the Mexican telenovela *Quinceañera*, but not telling Veronique about it. Each episode always started with a close-up of the teenage heroine's face. Oh, how I wanted to have her beautiful features, even though part of me laughed at the stupid plot. She cried as the music played. And as she looked down with her long eyelashes, the light reflected on her pearly lipstick. The song went something like: *Now the woman that was asleep inside of me awakens and, little by little, the girl dies.*

I wanted the girl in me to die as quickly as possible, and I will always be grateful to my friend Veronique for setting me up with Angel. He definitely helped me get close to what I wanted, but I still wonder why he just never called again.

Back in my Brooklyn apartment while my baby boy takes his nap, the first thing I do is Google him. I open the shiny white screen and I type his heavenly name, followed by his unusual last name. It's an interesting list of links. The first hits are from the State Department's Most Wanted list, followed by the Colombian government's list of the largest money-laundering businesses in the country. I'm left wondering: Was his father, the notorious mafioso after whom Angel was named, being extradited at the time? Is this why he didn't call me? Perhaps he noticed when I stepped on his foot one too many times during the more complex merengue turns? Did he find another girl to kiss, perhaps one without braces? Or maybe he just didn't want to bother.

I'll never know, I guess, but I can't think of him and not remember the Brut cologne ad campaign of the time: "Brut: For the man who doesn't need to bother."

That is a loose translation. The real commercial had a sexy woman's voice-over: "Brut: For the man who doesn't need to make too much of an effort." The sentence was pronounced with the speaker savoring every word, while a woman's hand slowly went under a man's shirt. His chest was hairy, and the hand had to navigate through the hairs and a thick gold chain. Angel always wore Brut and, needless to say, he never had to make too much of an effort with me.

Unlike him, I had to make great efforts at fourteen, though some had begun to pay off: Veronique and I had started riding around town in a black Fiat with red spoilers, driven by a chubby guy named Bula. His father shared his same round build and was a well-known politician, often quoted in the news. The Fiat's co-pilot was better looking and his name was Pipe, short for Felipe. Veronique and Pipe had something going on, so that secured us daily joy rides for a while.

It was Veronique's fling with Pipe that thankfully excused us from being considered *gasolineras*, or "gasoline moochers." Those were the girls who only went out with boys because they wanted to ride in their cars. Bula and Pipe always expressed to us their disapproval of *gasolineras*. However, though it was bad to be called a *gasolinera*, you reached rock bottom if you were considered a *zorra*, or "fox girl." And these guys also expressed their strongest disapproval for *zorras*.

From what I could understand at the time, a *zorra* was an easy girl. She was easily kissed and easily fooled, meaning, I guess, she kissed a guy before having secured a "relationship" with him. In order to let a guy kiss you, you had to be in the receiving end of a love declaration and become his *novia* officially, meaning you would receive phone calls and visits, and you would be seen around together. *Novia*, incidentally, means "bride" and also "girl-friend." I always wondered how one could have boys and fun rides without being labeled, or worse, being called a *zorra-gasolinera*. It seemed that Veronique understood all of these subtleties.

Working the right look was also a matter of degrees. You needed to look cute and sexy, but not like a *zorra*. In the weeks spent hanging out with Bula and Pipe, I was content with the joy rides, my new look, and leaving the complications of a love life to my friend. Until I met Angel.

Veronique and I first met him in our building's indoor swimming pool. We were taking a swim when we saw Angel and his gang. She had said one of her "things," something a little alluring, opening up the opportunity to hang out and maybe to go on a few rides, but without conveying we were the *zorra-gasolinera* type.

She introduced Angel and me and we all went up to my apartment. We sat around in my bedroom where my sister's bed sat close to mine. Angel sat next to me. I said nothing and neither did he. We just listened to Veronique's endless chatter and to the jokes his friends made.

In those few minutes, something had happened. I felt this lightness, this delight I hadn't known before. Sometimes we looked at each other as if we shared a secret, as if we stopped listening to the others and just pretended to pay attention to their prattle in order to continue sitting next to each other. For the first time, I felt like Veronique's subtlety was mine, and hers was gone.

After hanging out in my apartment for a while, we decided to go outside again. Angel's friends were being very loud and my parents were about to get home. Next to the swimming pool, there was a cement tennis court surrounded by eucalyptus trees, with roots breaking through the cement.

"Let's all sit over there," said Veronique.

"So where does your last name come from?" said an elfin guy to Veronique.

"My father is French," she snapped back, and I could see a faint giggle going around the gang.

"The French can't play any sports," said the slightly taller one of the crew, revealing a faint shadow above his upper lip as he smiled.

"It's true," I said, "we go to the French school and all we have to play sports on are cement patios."

"We're bad at soccer, but great in basketball," said Veronique, exasperated at my excessive humility.

"Angel's dad owns two soccer teams. One in Bogotá and one in Cali," the guy with the pubescent moustache offered in return.

I looked at Angel, who just blushed and looked down with a smile, as he caressed his jaw with his hand. *Brut, man, no effort*— the words just rushed through my mind. He had such a great shade of a beard at only sixteen. His skin was fair, his eyes green, and his hair was very black.

Then, recovering from my stupor, I said to Veronique: "Soccer teams? I didn't know you could own them."

She gave me another one of her sideways glances, and Angel, noticing the tension, said, "You guys should come and hang out with us at the soccer club sometime. It's fun."

That sounded good to both of us, especially because the rides with Pipe and Bula had started to become tense. We had made a bonfire a few days before on the tennis court and Veronique and Pipe, having had too much cheap apple wine, got into a fight, while my conversation with Bula had reach an all-time low.

When it was time to go, we walked Angel to his car, a red Cherokee jeep driven by two large and intimidating men in tight suits who came to pick him up. He held my hand part of the way. "You are going with Angel on Saturday," were Veronique's only words as we stood in the lobby, watching them drive away. Just like that, she had managed to arrange dates for us for the quinceañera.

"Can we take dates? I didn't think we were supposed to," I said.

"Who cares?" she snapped, almost yelling.

I didn't mind; I was filled with gratitude.

The quinceañera's name was Magdalena, like Colombia's most important river, which winds through the country northward and was, since the Conquest, the main artery for commerce, food, and people to the Atlantic Ocean.

Unlike the river, too shallow in some places and governed by perilous currents in others, Magdalena was a girl you could count on. Neither Veronique nor I had ever really talked to her before, but she had invited the entire class to the party. She was shy and unassuming, but the rumor was her father was throwing the house out the window for the party. She belonged to a clique that we referred to as "the pears," shy and studious girls with large hips.

The week leading up to the party, we spent our afternoons at Angel's soccer club in the outskirts of Bogotá. Our joy rides had just gotten a lot more interesting. Angel didn't drive, although he was sixteen already. His large and quite serious chauffeurs always did it for him.

At the club we would sunbathe in the soccer fields and Angel would treat us to grilled cheese sandwiches. It always felt like nighttime at the club's restaurant, with its velvet shades permanently pulled down.

"Do you know that this club also has a runway?" Veronique asked me out of the blue. "Yeah, right," I said, and she rolled her eyes back at me for the umpteenth time. I wondered if the planes landed there at night because I hadn't seen any yet during the day.

We could tell that Angel was both feared and respected at the club. The waiters, clad in grease-covered tuxedos, their shiny faces pale from lack of sleep, would fly to our table to serve us our snacks the second we'd sit down. I guessed it was no small thing to have a dad who owned two soccer teams. Angel, however, acted unassuming in the club, always thanking the staff profusely and trying to keep his loud band of friends under control.

During those afternoons, Angel and I would sit side by side and he would always hold my hand when we walked toward his car to see us off. And along with my teachers and parents, Veronique had also noticed my absentmindedness since I had met Angel. "You are definitely in the clouds. Do you even *know* what you're wearing for the party?"

We went back up to my apartment and I showed her the outfit. She seemed to approve. She didn't make a big deal out of hers, or of the fact that she was going to attend the party with Angel's friends, the loud gang of thinly mustached gnomes. I started to suspect she was desperate to try to make Pipe jealous.

When Saturday finally arrived, I couldn't believe my good luck. I teased my hair until it finally obeyed my command, and applied my makeup like the quinceañera from the telenovela: heavy eyeliner and pearly lipstick. I buttoned my blouse all the way up and secured the top button with the shiny heart brooch. I saw the pineapples, coffee beans, and bananas pouring down from my quarterback shoulder pads like a bounty. It reminded me of our national shield, its two golden cornucopias side by side, one spilling golden coins, and the other, tropical fruit.

Veronique's outfit was similar in shape: wide shoulders and very skinny legs. Her shirt was a plain purple satin that she wore with an oversized bow covering her left hip like an enormous, dormant moth. Like mine, it was buttoned all the way up and held together by a pin in the shape of a sea urchin. Her skirt was also a *falda chicle*, or "chewing gum skirt," which was sort of like a skirt made out of legging material. We both wore flat Mary Janes and thick pearly pantyhose to match our lipstick.

We had arranged to get picked up at Veronique's house and

when the buzzer finally rang, the doorman announced, "*Un señor Angel* is looking for *Señorita* Constanza and *Señorita* Veronique."

"Pipe is going to have a fit," yelled a euphoric Veronique, getting into the elevator and punching the number one with repeated violence.

Angel was waiting for us in the lobby. He casually kissed Veronique hello and then came over to me and gave me a slightly longer kiss on the cheek, slightly closer to the mouth: It's called a sidewalk kiss in Colombia. He looked beautiful in his white suit, his jet-black hair wet and pulled back. He was sporting a diamond pin of the Eiffel Tower on his lapel. How thoughtful of him, since Magdalena's party was held at the French Alliance.

We held hands until we boarded Angel's red Cherokee. When his bodyguards opened the doors for us, we were instantly hit with the musky scent of Brut. The bodyguards, looking freshly shaven (even those extra hairs on their thick necks), rode in the front. Angel, Veronique, and I were in the backseat, and Angel's friends were all in the trunk. They were already drunk and louder than ever. Veronique joined in the banter right away and I said nothing. Angel, noticing how nervous I was, whispered in my ear:

"I can't stand them either." I was already drunk with his scent when one of Angel's bodyguards offered us a bottle of Aguardiente. Angel took the bottle from the bodyguard so I wouldn't have to, and with a smile, he begged me to take a sip.

"It's not so bad, I promise. You just have to have a sip of soda after."

I tried it. My heart pounded as the strong anise flavor burned a trail down my throat. As we flew past the hills, Angel casually placed his hand on my knee. When I turned to look at Veronique, she made fun of Angel's outfit in French. "All in white. In Bogotá! People are going to flip!"

I knew what she was thinking as a fellow *cachaca*, a native of Bogotá: she thought he looked like a *costeño*, someone from the hot cities on the coast or a *calentano*, someone from the tropical provinces who hadn't taken into account that they were now 9,000 feet above sea level.

I just giggled back at Veronique mainly in an effort to make her shut up. I couldn't care less what people would say about him. As far as I was concerned, he could totally pull it off, even in our cold, Catholic, and modest moored city where everyone has always preferred shades of gray.

When we arrived at the party, I avoided saying hello to my other school friends. Magdalena's father greeted us courteously and indicated that we needed to make a "row of honor" for her. So we all stood there, very still, like the cranes that border the riverbanks at dawn. Magdalena was as unassuming as she always had been, even as she came floating in on her fluffy dress wearing heavy golden jewelry and lots of makeup in pastel tones. Her hair was up in a bun, except for some curls adorning the sides of her face and an uncharacteristic crest of teased bangs on her forehead. Her makeup made her lips fuller, her eyelashes longer, and her chubby cheeks somehow leaner. Rumor was she had bought her dress at Princesitas 2000, the finest store in Bogotá for quinceañera dresses.

Magdalena's father came behind her in the procession. He was wearing a tuxedo and patent leather shoes, just like her three little brothers who followed next. Her mother walked proudly, holding the smallest boy, who was just learning how to walk, by the hand. As Magdalena passed in front of us, Angel squeezed my hand again, and smiled without looking my way. I wondered if I had already become his *novia*.

The girlfriends I had had at school prior to Veronique, good girls from wealthy Catholic families, hardly spoke to me that night. And when they did it was to give me a quick "Hello" or to ask "Who's *that* you brought to the party?" or "He's not from around here, is he?" Even in their Madonna-inspired outfits bolstered by the eighties music roaring in the background, they still stood out as aristocratic Bogotá girls. "God, even the baby brother is wearing a tuxedo," I heard one whisper to the other.

The rest of the night I can sum up as a long, single kiss. On the dance floor, Angel and I went at it through most of the merengues, salsas, and American pop songs, only pausing to try certain steps and turns. I could almost read my friends' lips saying *zorra*, but I was too engulfed in Angel's embrace to really care. If I had become the queen of *zorra-gasolineras*, I might as well enjoy it.

Our only interruption came from a series of small explosions that almost brought the party to an end. I spotted Veronique chasing after Angel's friends like an angry chaperone, and Pipe laughing. The gnomes were blowing up the balloons with their cigarettes.

"I'm so sorry," said Angel. "Sometimes they can be real pains. Do you want me to talk to my chauffeurs? Maybe they can threaten to take them home."

"It's okay. I think Magdalena's father has made them stop."

He brought me closer to his chest again. "Your heart is pounding!"

I blushed. A new merengue started, and Angel decided to do some complicated turns on me to get my mind off of the balloon fiasco.

After the cake and pictures came the biggest surprise. They had hired a *minithèque* complete with neon lights. The effect of the neon lights made everything disappear, except for what was white. And the party became a devilish carnival of teeth, eyeballs and bras. Angel and I continued holding each other, oblivious to the change in lighting, which made all of Magdalena's family disappear, except for the little ruffled chests of her younger brothers, who zipped all around the dance floor.

Not one single part of Angel had disappeared. His white suit glowed and engulfed me in a bluish halogen light. My white shirt fused with his, and only the pineapples, coconuts, and coffee beans drawn in black lines vanished into the glow.

When the lights came back on, Veronique came to meet us. She was visibly annoyed. Angel's bodyguards were getting impatient and threatening Veronique to come and get us from the party. According to them, it was time to leave.

"Let's go," said Angel. "I don't want my father angry at me for keeping them too late."

We all climbed into the red Cherokee again. Veronique, Angel, and I took our places in the backseat, and his friends went back into the trunk. The bodyguards took the long way home through the hilly roads skimming the Andes. As the tires screeched in front of my building, and before I could say *"muchas gracias"* or *"buenas noches"* to his drunken gunmen, Angel asked me if I wanted to be his *novia*, and I said *sí*.

We rode the elevator together, gaining altitude, as the smell of his Brut cologne intensified. As we kissed, I looked up at the little purple numbers above the elevator doors become illuminated and then fade, dreading seeing the sixteenth floor light up. When we arrived, I stepped out of the elevator and turned around to face him.

"Okay, I'll see you," I said.

"Thank you, that was fun. I'll call you this week."

"Okay. Oh . . ."

The elevator doors closed before I could tell him I had seen him standing on the corner of 7th Avenue and 94th Street waiting for his school bus. I was going to suggest that I could wave to him the next time I passed by. Instead, I stood listening to the sound of the elevator car bringing him back down.

Walking to my door, I felt tipsy but I was still wide-awake. I turned the key and quietly tiptoed into my apartment. "Mom, I'm home. Going to sleep now!" I called out before locking myself up in the bathroom.

I had to untangle my hair for a long time. It had solidified into a painful, lacquered knot. I took a hot shower and tiptoed into the bedroom, where my seven-year-old sister was sleeping peacefully. I slipped into an embroidered cotton nightshirt, which waited for me under my pillow. It had been ironed to perfection. I remember feeling an inexplicable sadness getting ready for bed. Somehow I knew life would go back to normal in the weeks to come.

And so it did. Months went by without a word from Angel. One day, as I rode the school bus on another one of those frigid Andean mornings, studying algebra in French (a skill I would never use again in life), I felt a pang in my chest when I saw one of those Cherokee jeeps drive by. Was it filled with Angel and his thugs? I'll never know.

As I hear the first cries coming from my baby's room, I quickly clear the history of my Google search and shut down the computer. Before I go to pick him up from his crib, I decide to make him a fruit salad. I mix in some banana, pineapple, and orange, thinking about how fruits taste so much better in my native country.

That's something I'll always miss.

Queens for a Day

The Year of Dreaming: A Tale of Two Quinceañeras

BY Fabiola Santiago

"Es mi niña bonita, con su carita de rosa,
es mi niña bonita, cada día más preciosa."

—from the traditional father-daughter
quinceañera song crooned by Spaniard
Tomás de San Julián to the rhythm of
flamenco guitars

Whenever I get lost in the labyrinth of my imperfect life, I go back to the days when I was a dreamer, a schemer, a survivor, and lucky for me, a writer. To reconnect with the most genuine part of me—my adolescent soul—all I have to do is turn to the pages of the diary I began writing three decades ago at fourteen as I neared the milestone birthday that would deliver me into *mujer* territory.

In snippets of secrets revealed through bad poetry—in the obsessive measurements of my breasts, my waist, my hips, in the meticulous way I noted how I spent my first paycheck—a life unfolds. My words, at times clumsily strung together like the homemade hems of my miniskirts, are full of clues to who I was at the brink of womanhood.

The year: 1974. A disgraced President Richard Nixon faced impeachment. Patty Hearst had been kidnapped. The Miami Dolphins, the greatest football team, headed to a second national championship title. Russian dancer Mikhail Baryshnikov defected and joined the American Ballet Theater. Roberta Flack swept the Grammys with "Killing Me Softly." Duke Ellington died.

For me, it was the year of dreaming, and back then, as it has always been throughout my life, the dream can be summed up in one quest: Freedom.

At fourteen, my life was tightly wrapped in a parental—and Cuban—straitjacket. I wanted to wear miniskirts and bikinis like my American and my more progressive Cuban-American friends. I yearned to go dancing in trendy six-inch heels to the latest weekend "Open House," where bands named Coke and Clouds experimented with the new Miami sound, a mix of conga drums and electric guitars. But I was smothered by too many prohibitions when it came to my passion for fashion (a one-piece swimming suit in the '70s stood out like an Eskimo coat in summer) and by too many taboos when it came to boys (I would be chaperoned until I went away to college). I was smothered by too much suffering for the lost *patria*, by too many weekends spent visiting the exiled friends from Cuba, nostalgia as the only form of entertainment, the never-ending hope of a return to the island the only antidote.

Only the promise of some breathing room attached to turning fifteen gave me reason to celebrate.

The magical turn of the calendar on my birthday—St. Patrick's Day to everyone else in my new homeland—would mark my official cultural transition "*de niña a mujer*," which in practical terms meant I could wear makeup and nail polish with abandon, instead of behind my parents' back. Once past the quinceañera mark, I might entertain the idea of going to dances, of saying yes to a date, albeit chaperoned. This was all clearly a compromise since what I really pined for was to climb on the back of my hippie neighbor's motorcycle and be as free as he.

Nowhere in my dreams was there a big party with fourteen couples waltzing, nor a princess-style dress and a tiara.

Mine were refugee-size dreams. We were too poor for the traditional debutante party. That part didn't matter much to me. Marking traditions was my mother's dream. Mine were already hyphenated. I thought the party and the dress tacky and the emotional and financial investment in one birthday overwhelming. And I didn't want to hang on to Old World traditions that reminded me of our island home. Being Cuban came with wounds of loss as fresh as a Western Union telegram announcing a death. It came with all that overtime work in the factory just to put a foreign roof over our heads and comfort food on the table.

Being Cuban had become so sad.

Being a carefree American offered more possibilities.

At the brink of fifteen, I was plotting my escape from poverty, from traditions that came hand in hand with limitations, from the suffocating grip of overprotective parents, whom I thought sad and old-fashioned, and who now I see, with the clarity of having lived life and become a parent, as heroic. My year of dreaming was for them another year of suffering in exile, and at no other time had the chasm between us been so wide.

Only a little hole in my heart betrayed my teenage bravado. But no one could see it. It is only revealed in the pages of my diary on an entry dated March 3, 1974, written in Spanish fourteen days before my Cinderella birthday, and in the bittersweet tears I would shed on that page-turning day.

My middle daughter, Marissa, turned fifteen in 1999. A disgraced President Bill Clinton faced impeachment. The doomsayers were predicting Armageddon for the new millennium. The Miami Dolphins no longer performed miracles, but the Florida Marlins had, winning the World Series two years earlier, and becoming the team to watch. Russian President Boris Yeltsin survived his impeachment hearings. Celine Dion swept the Grammys with "My Heart Will Go On." John F. Kennedy Jr. died.

"Mom, what are we doing for my quinces?" Marissa asked one day without any previous warning.

"Your what?"

"My fifteens—you know, the dress, the party, the pictures."

"Quince," I corrected her. "Otherwise it's a literal translation of fifteens and it's wrong."

"Whatever."

More than correcting grammar, I was buying myself time, as I so often do when my three daughters exercise their amazing capacity to surprise me even though we keep few secrets from one another, and now that we are all grownups, act more like roommates and friends than the traditional mother and daughters.

It took me days to process Marissa's request to celebrate her birthday with all the trimmings of a traditional quinceañera. I dreaded the thought of the expense, but more than that, I still thought the ritual tacky, ridiculous, wasteful. Not a lot had changed from my own days as a quinceañera. My hyphenation had become an inextricable part of who I was—a Cuban-American woman, divorced and struggling to balance a high-flying career as a journalist while raising three daughters to become independent and well-educated women. Upholding Old World traditions still was not high on my priority list. I never suspected that it would be for the next generation.

What could possibly have come over Marissa, the most unassuming and inwardly focused of my daughters—the swimmer, the baseball player, the saxophonist, the pianist, the mathematician—to want to wear taffeta and come out of a pink shell in public?

It couldn't be the lure of the fairy tale dress.

While her sisters loved shopping and fashion, I had to drag Marissa to back-to-school sales, and at the end, it was always me who ended up picking a new version of the same thing she always wore: preppie Ralph Lauren and Tommy Hilfiger plaid shirts, jeans, and sweatshirts in nothing but blues, blacks, grays, and olives.

It couldn't be the debut dance with a boy.

She had a boyfriend. Her sisters had boyfriends. I allowed my daughters to have one as soon as they wanted to, and boys were like girlfriends at our house, simply part of the family picture. Marissa danced with her boyfriend whenever she wanted. In fact, they made music together in the school's marching and jazz bands.

The mystery behind her motives remained until I brought up the topic as I began writing about our quince memories, seven years later. But back then, the party, the dress, and the studio pictures in the dress were on, despite my reservations and her inability to verbalize why she wanted to be a traditional quinceañera so badly.

There was no talking Marissa out of any part of it.

"It's what I want," she had simply decided.

My mother was ecstatic. After practically shoving me into the quinceañera dress for pictures at a photo studio and Miami Beach's Japanese Gardens months after I had turned fifteen, and then having gotten a resounding no from her first granddaughter to the quinceañera tradition two decades later, she finally had Marissa, a quinceañera with a girlish heart like hers.

My first daughter, Tanya, has a flower-child soul like mine. For her fifteenth birthday, we went on a family trip to New York to soak up culture, and her most treasured gift was seeing *Les Misérables* in great seats. Much to my mother's disappointment, Tanya's only "quince" photos are of her in a flannel shirt acting up in Times Square and ice skating in Central Park—my kind of quince, and Tanya's.

With Marissa, I had no choice but to keep my opinions to myself, fork out the money for the dress, the photos, and the party, and go along for the real quinceañera ride.

It was a spectacle.

Here we were at Acosta Studio in Hialeah, the most Cuban city outside of Havana, and Marissa was trying on with great relish more white taffeta gowns than I had ever seen in my entire life. So many girls like her were doing the same thing, craving a return to tradition and embracing their roots. I read all about the trend later in the newspaper.

"Perhaps something simpler," I said to the seamstress shuttling between fitting rooms and helping us pick a dress.

My tomboy Marissa, I thought, couldn't possibly choose from this heap of lace.

"No, I like this one," Marissa finally said looking at herself in the full-body mirror.

I was stunned.

She picked the one with the most ruffles—puffs and puffs of them on the sleeves and all around the bottom of her umbrella-shaped dress. The gown came with lace gloves, a white cape bordered in white feathers, and a tiara just as grand. All I could do was give her my strand of pearls and buy her white satin pumps to match.

And so, my middle daughter would teach me all about the sweeter side of being a quinceañera.

My last memory of Cuba is the view from the window of our propeller plane as it took off from the airport from Varadero, the famous beach town. We were headed for Miami and, as it turned out, an eternal exile. As the plane flew over the lush, deep green landscape—a tropical menagerie of royal palms, plantain trees, and ceibas made the more beautiful by the aqua-hued coastline of endless white beaches—tears rolled down my cheeks despite my best efforts to contain them. My fortysomething parents, my eight-year-old brother Jorge, and I, ten years old, were leaving everything and everyone we loved behind, trading it all in for a Freedom Flight on a clear day in October of 1969. Two hundred and fifty thousand other Cubans did the same between 1965 and 1971.

As wretched as life was in Cuba, it was my paradise.

I didn't want to leave my beloved grandmother, who was always at my side. Or my cousins who were like brothers and sisters, my circle of best friends, and the boy I loved (dreamers are always in love). But as it was for many Cubans who did not embrace the Communist turn of the Cuban Revolution, the 1960s were for my family a time of instability, fear, and hardship.

Even though I was a child, no one had to tell me that danger lurked at every misstep, in every misspoken word. I sensed it in the militant vocabulary adopted in school, in the constantly un-nerved behavior of the adults. Behind closed doors at home, I heard whispered stories about people who had been executed be-cause they were deemed "traitors to the homeland," stories about people who were sent to jail for showing their disapproval of the government. My parents warned my brother and me to be careful with what we said, even among people who had once been good friends, even among family. Neighbors and relatives had turned into watchdogs, reporting each other to "save" the hard-fought Revolution from attack by "the imperialist enemy to the North."

It was never clear what we were allowed to talk about and what we were not. Any indiscretion, no matter how small, imaginary or real—getting extra toys besides the ones allowed in our ration book, eating steak for dinner, plotting our escape from Cuba—could get our parents in trouble.

"*¡No hablen de eso!*" my mother would scold us midsentence.

Other times, she would deliver the all-time favorite gag order of Cuban parents: "*Los niños hablan cuando las gallinas mean.*" Children speak when hens pee.

Uncertainty and fear became part of our daily routines.

From my bed, unable to sleep, I could see through the thin tulle of my mosquito net my mother sitting on a rocking chair and cry-ing as she waited for my father to come home. I did not know this then, but she had good reason to expect the worst. My father had been secretly helping antigovernment insurgents in the central Es-cambray Mountains by transporting people to safe havens and getting food and survival supplies to them.

One day, my father stopped coming home from work. He had been sent by the government to "*la agricultura*" forced labor in the agricultural fields. We had declared our intentions to leave the country and it was the price he had to pay for our freedom. My father's thriving small business as a flour goods distributor in our northern seaside city of Matanzas had been confiscated by the government in 1965. The olive-clad officials who took it asked him to stay on as an employee of the state. But my father proudly declined. Without a doubt, it was one of the most difficult moments of his life, one that he still recalls over and over, tears welling in his shriveled eighty-three-year-old eyes, as if this had just happened yesterday.

After he refused to work for the state and requested a visa to join a brother in the United States, my father was banished to *la agricultura*. It did not matter that he had to undergo an operation to remove kidney stones, nor that the incision, eerily wrapped around his entire right side in a red line, was still fresh. After he was discharged from the hospital, my father was sent right back to the fields, where he toiled from dawn to dusk, picking tobacco leaves and cutting sugarcane until we were finally allowed to leave.

My mother also had lost the job she treasured. A respected elementary school teacher who in her spare time taught illiterate young people and adults how to read and type, Mami resigned after teachers were required to add Communist dogma to their lesson plans. Without her career, my mother's full-time job consisted of insuring our survival in a society in which with each sunrise, official decree by official decree, the lives of people worsened.

She stood in long lines for rationed bread and rice, and she tried to get additional food supplies from relatives who lived in the countryside or by buying food on the black market. She tapped every possible friendly contact in the bodegas and the butcher shops to keep us well-fed and to send provisions to my father in *la jaba*.

Most of all, Mami monitored our lessons to insure that Jorge and I were not brainwashed by militant rhetoric in school. This street-smart role was not an easy one for my mother to assume, being an obedient and tradition-bound woman who, as the youngest of eight brothers and sisters, went from living under the strict rules of an overprotective matriarch to living under an overprotective, dominant husband. But with my father gone and our household identified as one of *"gusanos,"* worms, as people leaving the country were branded, it was up to her to shield us against the blossoming politics of oppression and indoctrination. It is hard for me now to imagine how my petite, polite, devout Catholic mother managed to stand up to the teacher who dragged my little brother by the ear to punish him for his "counterrevolutionary behavior." But she did.

Despite the hardships, my memory of my Cuban childhood is more fanciful than painful. Even though I knew what my parents were going through, I did not feel the pain and the fear with the all-consuming intensity of an adult. In childhood, the thrill of play and discovery is a more powerful emotion.

But for my parents, exile was like a blunt cut through a limb.

In exile, my mother cried every day for months. When she wasn't crying, she was sewing in a factory by day, at home by night, and on weekends, getting paid by the piece—five, ten cents—to attach a sleeve to a dress, a neck to a blouse. She was a great cook, but a terrible seamstress. She loathed sewing. In Cuba, she either bought her clothes at El Encanto or had ours made by the best seamstress in our neighborhood, Ofelia, who lived right next door. In exile, she had no choice but to take the only available jobs in clothing factories. My father, who worked in a window-making factory, often reminded her that this was all temporary. The totalitarian regime would surely be deposed and we would go home, he would promise, the torch bearer of hope in our family to this day.

After almost three years of exile and no miraculous return in sight, my parents poured every penny of their overtime savings into a new house way out in Miami's suburbs. I translated the mortgage documents and the closing transaction. Only when I bought my own house and had to hire a lawyer did I realize the enormous responsibility of what I had been asked to do back then.

Our little piece of America was a sandy lot with nothing but the newly built three-bedroom, one-bath house on it. Jorge and my father attached a Cuban flag to a pole and raised it in the backyard.

"*Primer territorio libre de América*," Jorge called our house, still parodying the slogans to which we had been subjected to in school in Cuba. My mother and I laughed, clapped, and captured the moment in a photograph.

Our new house was a huge improvement over living in a cramped apartment or in the old rental house by the airport, where it felt like the airplanes were going to land on our roof.

The new neighbor to our right was an avid reader and she passed on to me all of her *Buenhogar* and *Vanidades* magazines, and I gobbled one Corín Tellado *novelita* after another. In another house down the block lived a Kris Kristofferson lookalike with eyes the color of caramel candy, a perennial suntan, a devil-made-me-do-it smile, and shoulder-length hair that blew against the wind when he rode his motorcycle. I was in lust. (Dreamers are always in lust.)

In school, I became "Fabby," a clue to who I had become, an adolescent desperately struggling to become as American as I could possibly be. My American teachers had nicknamed me—and they pronounced it *Faye-bee*—after they tired of coming upon my name every day in roll calls packed with Johnsons and Smiths, and a sprinkle of the easier Perezes and Garcias. I became a founding member of the school newspaper, the *Hialeah Junior High Highlights*; it was the birth of a lifelong career. I wrote a gossip column, a song dedications column, and a "Dear Fabby" advice column (students were asked to slip their letters through the vents in my locker). My English teacher, Mrs. Rosenberg, egged me on to keep writing. At the top of one of my compositions on love, she wrote: "Someday I am going to buy your books."

It was during my last year of junior high school that I began writing bits and pieces about my life in my diary. Thanks to my meticulous note-taking and my passionate prose on matters of the heart, I remember the years from fourteen through eighteen with a clarity I cannot muster for other periods of my life.

Long ago, I lost the diary's tiny key. To pry into my old secrets, I have to break into it like a thief or a nosy mother. But it's not hard. The tiny blue lock easily yields to the tip of my tweezers, and with just a light twist, I hear it click open. It makes me wonder how many times my mother broke into my diary all those years, although I doubt she ever did. Because if she had, even three decades later, we wouldn't be on speaking terms.

Such were my secrets.

"His hands so near me . . ."

"They can't understand that I am a person and not their private property."

"They depend on me for everything."

"I want to be able to choose for myself what I want to do."

And this prophetic line: "Someday I hope to write to the whole world."

Its plastic black cover is speckled with beige nail polish, a testament to my devotion to both doing my nails and writing down my feelings. On the outside, in groovy white letters that date my youth, the thick journal proclaims "D-I-A-R-Y," and inside each of those letters, I personalized it in blue ink: "F-A-B-B-Y." Years later, with the confidence of education and after reconnecting with my roots through a journalism career and the building of my own family, I would stick to another version of my nickname, spelled the Spanish way, Fabi, and when my byline became professional, nothing but the real thing would do.

But first, there was assimilation.

To survive my Old World parents, I went underground.

I made my own miniskirts by cutting and sewing a new hem on the hand-me-down dresses I got from a friend of my mother's at the factory. I would wear the minidress like a shirt tucked into a regular skirt for the ride to school with my father, then take the longer skirt off as soon as I found a bathroom. Or I would roll my own knee-length skirts up at the waist until they became minis. Once I had my image the way I wanted it, I pursued my motorcyclist love interest next door. I needed some kind of cover to get close to him, so I began tutoring his little sister in reading and writing.

I plotted my escape route from home: Make the grades, get rid of the accent, get a job, save the money for college, go away to college. I was already making the grades, going from English as a Second Language programs to mainstream honors classes. And as that fifteenth milestone loomed, I wanted, more than anything else, a job. Children weren't supposed to work until they turned sixteen, but my parents had agreed to let me work part-time after I turned fifteen, the magical age of maturity in the Cuban calendar. At least *that* worked to my advantage. As young as I was, I instinctively knew that making my own money would buy me independence—not to mention a great pair of shoes.

I wrote about all this in my diary, chronicling the dreams, the gains, the losses, the heartbreak during those months approaching my fifteen and beyond. It's surprising to read how candid I was about some of the most intimate details of my life. In the pages of my diary, I was free. And I issued a warning to those foolish enough to violate my privacy. The first page declares my intentions about writing "the things I felt like saying but didn't, things I feel or felt were important and somehow secret." It asks the person who finds the diary to please "burn it without reading, if that is possible to ask of humans."

After all that gravity comes the fun stuff—my measurements: 35, 28½, 35. My weight: 113 pounds. Seven days before my fifteenth birthday, I was obsessed with thinning my waist and losing the extra pounds that our nightly Cuban dinners of rice and beans and *bistec/picadillo/ropa vieja*, and my American school lunches of hamburgers/pizza/hot dogs and pound cakes for dessert, had put on me. I lied about my height, or maybe in my dreams I was taller: 5'1". Truth is that I have always been five feet flat.

Soon after my quince, I noted that on the *Buenhogar* diet, I managed to bring my weight down to 111. I also got my first job as a salesgirl in the Cuban baby clothes *canastilla* and lace-by-the-yard store, House of Notions. I lied and told the owners that I was sixteen. Nobody asked for papers then. Starting pay: $1.75 an hour.

In my first paycheck, I noted in my diary, I earned $42 for twenty-four hours of work, but $7.26 went to taxes. I put $5 in savings, bought $20.80 shoes for myself, and put another $5 aside to buy my brother shoes with my next paycheck. When all was said and done, I had 65 cents left for the week's expenses. I also noted that somehow I had lost track of $2.29. Later entries show that during the summer, I worked forty-plus- and fifty-plus-hour weeks, and that with every paycheck, I bought clothes and shoes for my family.

There is not one word in my diary about my quince, nor about my mixed feelings about being a quinceañera. My father was very sick and some things are so painful and scary we cannot write about them as they're unfolding. My father was often in physical pain. He was diagnosed with kidney stones once more, and he faced another operation to remove them. There was talk about him losing a kidney. It was a time of fear for all of us, that familiar uncertainty of past times making a comeback. He was our primary breadwinner and we didn't have a lot of family to fall back on. Perhaps it was this reality that led my overprotective father to allow me to work that summer.

His operation was in February, my birthday in March. I could not dare to dream about what I wanted for my quince—a simple party with my friends from school and the boy next door. Nor could my mother dream about what she wanted for her daughter—the fancy white princess dress, the studio pictures, the fourteen couples doing the waltz, Tomás de San Julián crooning "*Es mi niña bonita*" while my father and I danced.

Given our circumstances, there was no talk about my birthday celebration beyond having a new dress made and celebrating the occasion with the usual Cuban cake *de merengue* shared with the closest of family, a ritual that marked every one of our birthdays.

Or so I thought.

I don't remember exactly what excuse my parents used to get me out of the house the day of my fifteenth birthday, but when I returned—"Surprise!"—all of my best friends were there too. They and their parents had conspired with my parents to arrange a surprise party. There was *cidra* on the table for a toast and a birthday cake adorned with a quinceañera doll. I remember that I was wearing shorts and everyone was all dressed up. Crying tears of joy, I ran to my room to change into the fancy dress my mother had made as a present.

It was a long blue spaghetti-strapped dress accented with fashionable blue feathers across the top. To match, I had gotten a pair of white platform heels, also as a present. They were humongous and made me look taller than my desired six inches. Someone brought me an orchid corsage. All my best school friends—Carmen, Clara, Irene, Mercy, Matilde, Laura, and Ileana—had pitched in to pay for the party and to send me an arrangement of carnations, in blue like my dress, my favorite color and instantly my party theme color. And the girls had brought along to the party enough boys to dance the night away.

From Cuba, two yellow Western Union telegrams had arrived, one from my cousin María Elena and another from Abuela Ramona. *"Muchas felicidades en tus quince años, cariños y besos de tu abuela."*

They had remembered.

I was the happiest of quinceañeras that night, even if my love interest, Mr. Motorcycle, never showed up to my party. I danced with his equally handsome brother instead. My mother took pictures. My twelve-year-old brother Jorge, in a new red jacket, flirted with the girls, popped into the picture when my friends were singing "Happy Birthday," and blew out the candles himself. I danced with my father, who looks a little pale and aged in the pictures but ended up healing quickly from his kidney operation without missing too many days of work. It was a special birthday indeed.

My mother, however, was not done marking the milestone.

Months after my real birthday, she announced that she had saved the money to get me studio photographs in a rented taffeta quinceañera dress. I didn't want the photos, and most certainly not the gaudy dress. But there was no getting out of it. I, however, came up with my own form of protest. I refused to wear makeup to the photo session. Wearing makeup had gotten old pretty quickly. Once granted permission to glamorize my face, the ritual lost its lure. No one ever said how great I looked in makeup. Everyone noted how naturally beautiful I was without it. It takes some women a lifetime to learn this, but at least this one lesson I learned at fifteen, and it has served me well into my forties.

And so, my mother dragged me to a photo studio and to please her I posed for the photographs. I chose a dress draped low on the shoulders—by then I had realized that my best assets were my C-cup breasts—and posed as instructed with fake red and white roses, a lace umbrella, and a royal blue cape. I wore long white gloves, a tiara, and the silver necklace with a heart my mother gave me as a present. When this was over, the misery prolonged itself all the way to the Japanese Gardens for outdoor pictures.

But today, as I cradle in my hands the old yellowing album labeled MIS QUINCE AÑOS, its worn pages so lovingly and creatively patched up throughout the years by my mother, I can only feel her love. I also can see with clarity in an off-guard pensive moment captured by the photographer, the contour of my dreams, the depth of my secret, revealed in a March 3 entry in my diary. It's a poem in Spanish entitled *"Dedicado a mi patria, Cuba."* A poem that rhymed like a child's, dedicated to my homeland, a poem in which I, too, dared to dream of a return.

> *". . . Cuba, mi patria linda,*
> *oh, por qué de ti partí,*
> *será que al igual que Martí*
> *quiero verte soberana . . ."*

As the next two years unfolded, I added more verses to the poem, confessing that I felt like "a coward" because I could not talk about Cuba without crying. How could I forget the place that "cradles in its bosom" someone I so love, I wrote about my grandmother.

"Muchos piensan que no te quiero
otros, que ya no te recuerdo,
pero yo te aseguro, Cuba,
que siempre te llevo muy dentro."

Breaking into my diary is a ritual I relish. When I was guiding my three daughters through the difficult teen years, it became a must to go back and pry into my thoughts and my girl-to-woman secrets. Reading about my angst over body image, family dynamics, the burdens of heritage, and those bad boys helped me remain close to my daughters' hearts.

Yet my musings didn't help much when it came to celebrating Marissa's dream quince. As much as I had tried to bridge the gap between our generations, my dreams and hers seemed to be worlds apart. I would have known just what to do if she had wanted to go to Paris. But when it came to throwing a big party with all the trimmings of tradition, I was out of my league.

That required a village, and the project became a family affair, unique to our spirited multicultural brood, and to this day, the most remembered of birthdays.

First, there was the issue of renting a banquet hall for the party. Marissa and I went to see close to five halls in our area—and were left arguing about the merits of renting a space, she in tears, I in fear. Horrified at the expense and the thought of the potential security risks of half the school showing up to the party, I decided to put on the brakes. No banquet hall; and it wasn't up for debate.

My parents came to our rescue and offered a compromise: their house. The little house had aged gracefully over the years. They had put in a fairly new, good-sized roofed patio, perfect for a bash. The earlier version had served me well for my quince surprise party. On that tilted concrete slab my friends and I had danced happily under the stars, as if we were celebrating in the fanciest of banquet halls. The refugee generation, I tried to explain to Marissa, had no choice but to make do with what we got. As the song said, "If you can't be with the one you love, honey, love the one you're with." She didn't seem to appreciate the lesson, but in the end, she embraced the party venue.

Then we went to work on the rest while family members pitched in to help. My mother and sister-in-law, gifted in arts and crafts, made decorations for the tables and party favors in pink and white. Marissa's godmother, a former television journalist, agreed to be the master of ceremonies using a script my mother wrote. I ordered a three-tiered cake, white with pink roses, and the traditional plastic quinceañera doll on top. We rented tables and chairs for guests—and for Marissa a fancy white wicker chair that, once decorated with tulle and ribbons, became the throne of a quinceañera.

Some dreams require tweaking to become reality—and reality becomes a better version of the dream. That's how it was for me. That's how it was for Marissa as she faced yet another challenge: the matter of braces. Marissa was still wearing a mouthful of them. We consulted the orthodontist about temporarily removing them, but that was impractical and costly. The solution: Marissa practiced the perfect half smile that hid them. It worked, and ironically, the soft smile and her gorgeous long brown hair draped down her back—au naturale—gave her photos the sweet moody look of innocence.

Anything else would have been fake. When I see Marissa's photos, compiled in a bride-white-and-gold-trimmed album that reads MIS QUINCE AÑOS, just as mine had, I see the vulnerable soul of the shyest of my daughters, quiet in her beauty yet comfortable on stage.

Her big day came on a January night, warm enough, as it typically is in South Florida, for bare shoulders and halter-top gowns.

It began with a trip to the beauty salon to style her hair into a princess-worthy bun. Instead of coming out of a pink shell for her debut, she walked out of the house, hand in hand with her father and me, to greet her guests. She took a seat in her place of honor, the chair-throne bursting with tulle and ribbons. Instead of waltzing couples, we came up with our own ceremony. Fifteen of her most dear relatives and friends brought Marissa a pink rose to mark all the years. Her little sister Erica sang for Marissa a lovely rendition of "God Must Have Spent a Little More Time on You."

To mark her transition, Marissa danced with her American father to one of his favorite pop songs, "Lady in Red," and with my father, her Cuban *abuelo*, to the traditional "*Es mi niña bonita*" by Tomás de San Julián.

We ended the pomp and circumstance with a barrage of emotional thank-yous, turning the night into the salsa and pop music party it was begging to be. While satisfied that my daughter had a good time, the mystery remained surrounding Marissa's surprising embrace of tradition.

Wedged in between two boisterous, opinionated sisters, Marissa has always been perceived as the understated one. She was the insecure one, at least on the surface, whose passion for music has always revealed another side to me. I had signed up all my daughters for piano lessons from the time they were five—the three were a charming trio during the annual recital—but the other two quit, and Marissa was the only one who kept at it, year after year. She also played the saxophone in her middle school and high school bands.

But from musical performer to quinceañera is quite a stretch, so I begged Marissa, now twenty-two and living in another city where she is studying computer technology and working a full-time job in retail, to tell me why she had wanted so much to celebrate her fifteenth birthday in taffeta and pearls.

She laughed her usual quiet laugh and was silent for a few seconds.

"It made me feel special," she finally said.

It is so simple a wish, but the stuff of which quinceañera dreams are made. And it is wishes and dreams that shape a life.

As for me, I still write in journals, love surprises, and miss Cuba. Sometimes, but only sometimes, I still dream about riding on the back of Kris Kristofferson's motorcycle. But now I prefer my reality: driving my convertible—top down, of course—against the wind.

It All Started with the Dress

BY Leila
Cobo-Hanlon

I had felt young and pretty that morning. But by the afternoon, as I stood in front of the four-way mirror at the dress shop and saw my replicated image from in front and behind, I felt ugly.

The seamstress and my mother watched expectantly. But I could tell the sight of me pained my mother. Her mouth was slightly open, her eyebrows raised ever so slightly; she was waiting for the shit to hit the fan. And then, it did.

"I look like a cow!" I wailed.

The dress was made out of vaporous emerald green chiffon. I had chosen the fabric and the cut to accommodate it, a Grace Kellyish strapless number that cinched at the waist then puffed out in waves of green, falling just above my ankles. Admittedly, it would have looked perfect on Grace Kelly or on some other tall, reed-thin goddess. But on a chubby fourteen-year-old with a broad back and a waist that had yet to fully form, it looked, well, ungainly.

"Mmm, no honey, you don't look like a cow," said my mother dubiously, reaching out to smooth an imaginary wrinkle.

"I look like a whale!" I cried before bursting into tears.

My mother did what she does in moments of crisis. She gave marching orders like the top general of her own imaginary army. She unzipped the dress from my sobbing back, told me to get dressed, handed it to the perplexed-looking seamstress and told her to take it away, and then hurried me out to the car.

We rode in silence for a while, the cooling general impassive in the driver's seat. "Leilita," my mother finally said, "can we just go to a store and buy a dress that's already made? A normal dress?"

I shook my head furiously. "Only if it's green," I answered and crossed my arms. My eyes were watering. In my little world, I was going through the most serious of calamities, with no solution in sight. I was less than a month away from my *fiesta de quince* and I still didn't have my dress.

In the beginning, my mother, the kind of woman who wears linen on an airplane and gets off without a wrinkle after a six-hour flight, thought this green dress notion was inconvenient albeit mildly amusing. But after scouring every single store in town without finding anything remotely green that fit the profile of a *fiesta de quince* dress, she was losing her patience fast.

"I am going to call your aunt and have her look for something in Bogotá," she finally said. "And then I'm asking Ruth to look for something in Miami," she added, referring to the owner of her favorite boutique, who regularly brought clothes in from the United States. "But," she interrupted when I started to protest, "I get to choose the dress, or you're stuck with that whale outfit. I mean it."

This is the thing: *Fiesta de quince* dresses, at least back then, were supposed to be white, or pink, or anything pastel and virginal. It was the mid-1980s in Cali, Colombia, and at the time, I would wager, most of us graduated immaculate from high school.

I was the youngest girl in my class, and when my fifteenth birthday rolled along, well into the tenth grade, not only had everybody else's *fiesta de quince* come and gone, but my friends now considered it terribly passé. It had become a slightly tacky tradition that belonged to the realm of the newly rich drug dealers and their growing dominance within the once-genteel Cali society.

"Who cares about the dress?" my friend Elisa asked the next day in school, the disdain in her voice making the term sound like a bucket of maggots. "I mean, who cares about a *fiesta de quince* in the first place? Does anyone even *do* that anymore?" Elisa was a good friend with a particularly mean streak that was usually directed toward others. But that day, apparently, it was my turn.

"If you think it's so stupid, don't go then," I answered crossly.

"Oh, I *have* to go," she said snidely. "My mom is making me. But I have plans to take off and go to Unoclub and then come back before she picks me up."

Unoclub, which shut down at one a.m., was the one nightclub parents allowed high school students to go to. At fourteen, I was absolutely forbidden to set foot in it, though most of my friends claimed to have made the pilgrimage there already. It was rumored to have a fearsome owner named Mirta who visited the tables of couples getting too close for comfort, turning up their table's individual lamp light and scolding them. Back at school on Monday mornings, my friends shared post-Unoclub bonding sessions that excluded the likes of me. Save for the occasional party, I was usually home on Saturday night, watching late-night TV. I understood Elisa's dilemma. If I had to choose, where would I go? The slightly illicit ambience of Unoclub or the gleaming brightness of my *fiesta de quince?* Obvious.

Ours was an American prep school, with small, tight-knit classes. Our entire grade was made up of sixty people, of which I—with my short, curly hair and thoroughly terrible aptitude for sports—wasn't the prettiest, or the most popular. But I was generous with my test answers, book reports, and school lunches and felt I had enough backing in each little subgroup to pull off a successful all-class party.

I had also very recently acquired my first boyfriend, a boy who wasn't from school, who wasn't a jerk, and who had proposed quietly during a party at his house. "*¿Nos cuadramos?*" he had whispered as we slow danced to the Bee Gees' "How Deep Is Your Love?" My heart had soared, because now I wouldn't be boyfriendless at fifteen, the equivalent of an old maid in Colombia. My *fiesta de quince* would be the place where all my friends would see that I had a guy by my side too.

I hadn't told a soul about the boyfriend, afraid that we'd be broken up by the party. Then I panicked over the very real possibility that the party itself would fall apart; that these same friends would come then leave in the middle, off to a place I had no access to. I went home with my stomach in knots. I had nightmares about trying on a parade of green dresses the night of my party and having none of them fit as the clock ticked steadily past eight, past nine, past midnight, past my *fiesta de quince.*

I can't really tell you why I was so set on having a *fiesta de quince*. But I had latched on to the idea of that party in the same way that years later I latched on to the idea of a big, traditional wedding. For whatever reason, I loved the notion of the evening gown, the pomp and circumstance, the waltz with my father, the fanfare. I knew I could ask for other, perhaps more glamorous things: A weekend in New York, a trip to Europe, a cruise, a nose job—the new must-have accessory of the season.

But for me, the party was as glamorous as it got. It was a chance to be a grownup princess at fifteen, in a city that now feared excess.

Cali had never been known for conspicuous consumption, but it was a city where names and standing mattered greatly. Wealth wasn't exactly flaunted, but it was most definitely there for the observing. When the kidnappings by leftist guerrillas started in the early '80s, fear dampened the exuberance somewhat. Then, the common criminals trampled on its luster some more. Stories abounded of thieves walking into homes during major parties and taking off with the gifts. That's why my *fiesta de quince* would be recorded for posterity by a picture in the newspaper *after* the fact, not before.

I preferred things the way they used to be. Not just because it was safer, but simply, because I found it more compelling. Raised on my parents' tales of elegant soirees and an endless supply of turn-of-the-century romance, I would have been delighted to be a 1940s debutante, filling out dance cards, sipping martinis, and wearing long gloves that would've hidden the fact that I bit my nails.

Only one other girl in my class, María Victoria García, had had a *fiesta de quince*. She was popular, traditional, and boring; the kind of girl who blow-dried her hair every day and aspired to marry quickly, marry rich, and have many babies. I remember she wore a pastel blue dress to her party. I dared to be different, even if it only meant defying a color scheme. I wore the brightest possible hue of green. It was my very favorite color.

There are pictures of me as a little girl wearing green dresses for the girls-only birthday parties we used to have that doubled as beauty pageants. My friends had to come dressed in floor-length gowns and parade in front of a panel of judges (my sister and her friends). We would crown a winner (it was never me) and had a total blast, even though the other moms would grumble for weeks beforehand about having to find evening gowns for their eight-year-olds. For my *fiesta de quince* I wanted to wear a long dress (long *green* dress, that is) all over again.

In retrospect, I can see how tricky it was to pull off a traditional bash when tradition was not in vogue anymore, when wearing a suit and a gown were considered an ordeal. I admit I sympathized with some of my party naysayers. After all, the last *fiesta de quince* I had gone to was completely against my will. My cousin Laura was visiting from Bogotá and jumped at the rare chance to party, even if it meant going to the home of someone we barely knew. To top it off, the quinceañera had been forced to invite us by her parents.

Have you ever been at a party where you know no one? Really, no one? That's what this was like. Except we weren't sophisticated twentysomething professionals yet who could strike up casual conversation, but terribly insecure fourteen-year-olds. I spent most of the evening near the buffet, nibbling unhappily and watching other girls get asked to dance. Even Laura, who had two left feet, got herself a partner; granted, it was for one miserly song, and he was pimply and ugly, but it sure beat my predicament— sitting alone on a stiff chair while pretending to be incredibly busy inspecting the tablecloth and slowly melting in mortification.

The next morning over breakfast I told my mom and sister, Roxi, about the party. I was a great gossip even then. "All the guests are there—all these people I don't know—just milling around drinking punch, not even real drinks—and the birthday girl is nowhere to be seen. Then all of a sudden, the strains of a string orchestra! It was Strauss's Viennese Waltz, the music *de rigueur* for all *fiestas de quince*. I looked around, searching for the party girl and her dad, thinking they've already taken over the dance floor. But no. People start looking up, *ooh*ing and *aah*ing, and there she is, at the top of the staircase with her long gown and her long, perfect hair, and her guests actually start clapping—clapping! As if she were a queen before her peasants or something—and she slowly walks down the stairs."

"It was the tackiest thing!" piped in Laura, only to then quickly wither under my glare. I was still pissed off at the fact that because of her I had to submit myself to four hours of staring at the floor. In a flash I remembered Elisa's odious little words about my own party: "I *have* to go. My mom is making me." I would hate to have people talking about my party with the same derision I was using to dump on my own experience of the night before.

"Mami," I said seriously. "My party has to be awesome. I don't want anyone having a bad time. And, please, please, please. It can't be tacky." Okay. So the word "tacky" is alien to my mother's genetic makeup. But *fiestas de quince* are, by definition, tacky. Think about it. It's a miniwedding, down to the cake, the flowers, and the gowns. Except you have all these nubile teenagers celebrating nothing more than the infinite sexual awareness of being fifteen. Forget that whole notion of the passage to womanhood, blah, blah, blah. I promise you, no one thinks about that when planning this party. Parents want to impress, and we want gifts, a good time, and covert make-out sessions on the dance floor.

"Hold me closer," my sister demanded as she went over the essential tips to dancing with an over-eager partner of the opposite

sex (played by me). I held her close with my left arm around her waist, and as I tightened my grip, up flew her elbows, smack between my chest and hers. I couldn't get any closer, couldn't even thrust my hips at hers.

"See?" she smiled triumphantly. "You don't even have to push him away. Just put the elbows up, he'll never know you're doing it on purpose."

That night my father went over the other fine points of dancing. Holding me firmly but respectfully around the waist, we went over the basic one-two-three waltz steps slowly. "Head up, back straight, and follow me, follow me," he sang happily as he twirled me around on the granite floor.

My father danced with elegance and flair, keenly aware that dancing was a couple's sport and that making his partner look good was tantamount. He could dance anything—big band, salsa, mambo, rumba, even the cha cha, which he had discovered in the discos of New York City as a young man. There was nothing embarrassing about dancing with my father or watching him dance. At parties, it was common for a circle of people to form around him and my mother, watching their moves.

But only after signing me up for dance lessons had my dad begun to ask me to dance at parties. And while we practiced that day I could see in his face how proud he was, not just of the moment, but of the fact that I was actually following his lead in a way that did both of us justice.

"Ricardo, your turn!" he called up to my brother, who, according to tradition, was also supposed to dance with me. "I'm not dancing!" Ricardo hollered back from upstairs. He was sixteen, a classical guitarist, and only wore black. I had never seen him dance in my life. My dad shook his head.

"Just this once," I said pleadingly. "Come on. One little dance!"

"No way," said Ricardo, incredulously. "This is so stupid. Why are we waltzing in Cali anyway?"

"Because I want gifts!" I shouted happily.

"Because we're celebrating your sister's coming-of-age," said my father sternly.

Then he turned all his focus on me. "Let's do this again," he said, clearly nonplussed that Ricardo wasn't going to join in but determined not to lose the spirit of the moment. He was, after all, spending a small fortune on this little fete, but more than that, he was making me happy. I can't imagine that the prospect of a hundred teenagers inside his house could have been a joyful one.

The week before my party, I lobbied furiously to ensure I had a quorum, cornering everybody I liked and asking them, "You *are* coming, right? *Right?*" Elisa had stopped talking about skipping out early and there actually was—dare I say it—a buzz surrounding the event. Despite these minor reassurances, my nerves were raw. I hadn't hosted a party since my last beauty pageant, and despite assurances that "Yes, yes, yes, we'll be there," I couldn't help thinking they wouldn't be. Like the unlikable character of a book I had recently read, I would be left sitting alone in my living room, surrounded by untouched mounds of food and apathetic waiters, crying on my green gown, which by that point, we had finally found.

It came from Miami via Ruth's boutique, and it wasn't emerald green but aquamarine green. It wasn't chiffony or strappy, but rather demure. It had a V neckline, an empress waist that flowed into a pleated skirt, and sleeves that fell just above my elbows. But the color still suited me, it still made my big brown eyes sparkle and my cheeks flush. And, in the world of *fiestas de quince*, it stood out like a shiny little emerald. Most important, I didn't look like a whale. Not even a cow. Instead, much to my mother's delight, I looked quite pretty in green.

The day of my party I woke up to our doorbell ringing. Flowers. Flowers poured in throughout the day, overflowing in every corner of the house. Their sole purpose was to celebrate me, to make me happy, to let me know how important I was. If I had known then that I wouldn't see such displays of affection again until my wedding day, perhaps I would have appreciated the smells and textures of all those bouquets a little more.

My cousin Laura from Bogotá arrived that morning too, carrying with her, wouldn't you know it, a green chiffon dress. I decided I wasn't going to let this ruin my day (although I felt like throttling her). I mean, what were the chances? Whatever. She also came sporting a gift from my aunt, a beautiful band of intertwined gold strands that I wear to this day.

Unlike many quinceañeras on their big day, I had no appointments at the beauty salon. Given my hair situation—unruly curly—and my loathing for makeup, there was really no point. Instead, I watched the activity unfold inside my house. Waiters strategically cleared entire areas of our house for dancing and for setting up tables. *Para los jovenes*, the living room and the garden areas. *Para los grandes*, the family room and the terrace.

This is the tricky part, you see: If you want a good party, meaning a party with tremendous gifts, you need to invite your family and your parents' friends. There is no elegant way out of this, otherwise you'd be stuck with a bunch of records or teddy bears as presents. Then again, you don't want a party full of just old farts. Nothing kills the ambience quicker than having *los viejos* watching you dance. Segregation was *absolutely* necessary once the waltz with my dad was over.

In our huge kitchen, my mother's caterer, Elisa, was busy

making the Lebanese food our house was famous for: Kibbeh and stuffed grape leaves alongside sfijas and pita bread. For dessert, baklava and graibes, the little butter cookies my classmates craved after school. And while the food was distinctly un-Colombian, the drinks were definitely in line with the culture. Which is to say, alcoholic drinks would be served, even for the fifteen-year-olds. Rum and coke for the boys, creamy Alexanders with cherries for the girls. No, this wasn't considered irresponsible back then. It still isn't. Fifteen-year-olds don't drive in Colombia, and in the controlled environment of a *fiesta de quince*, where a waiter personally poured the shots, getting drunk was unheard of. At least in my world.

By 7 p.m., everything was ready, and so was I. Was I the prettiest I had ever been? Not sure about that, but I certainly felt the most special I had ever felt. As I walked down the stairs in my dress (before the guests arrived, mind you. I wasn't going to have any of that princess b.s. from that other *fiesta de quince*), I surveyed a house that had been decked out entirely for my enjoyment and pleasure. It was a strange feeling, perhaps one of too much importance for a fifteen-year-old, except that my parents' faces reflected utter confidence. Tonight I could do no wrong.

I paced nervously downstairs, even though it was early and no one was slated to be there for at least another hour. We shot the family pictures; Roxi beaming, Ricardo pouting, my parents smiling proudly, my grandmother—the haughty doyenne—looking remarkably pleased with herself and her youngest grandchild.

When the doorbell rang finally, it was only the newspaper photographer, who came in, took a perfunctory picture of the family, and left in a rush, perhaps for another *fiesta de quince* that had to make Monday's paper. Embracing the importance of the situation, my grandmother had given me a "good" piece of jewelry, a huge emerald ring, Colombia's precious stone, to match my dress. It was a woman's ring, an emerald set high on a round pile of tiny diamonds, that years later would be referred to by my sister and I affectionately as the "birthday cake ring" because it resembled the tall, layered designs of our *fiesta de quince* concoctions. I remember it felt heavy on my finger as I twisted it round and round, waiting for someone, anyone, to arrive.

By 9 p.m. I started to panic. It couldn't be, I thought. But it was. Nine in the evening and not a soul. And then, the doorbell rang again. But this time, it was my boyfriend, who wasn't a jerk, who wore a dark suit and tie and looked better than I ever remembered him looking. He had brought with him a long jewelry box containing a slim, gold bracelet that I put on right there and then. Now I had two gold bracelets and a ring on. I had gone from demure to nouveau riche in a matter of hours.

As if on cue, a stream of people started coming in. Two hours late, but very on time in Colombia, after all. It was everyone I thought would come, everyone I hoped would come, even those people I thought would never, ever find it in their beings to set foot in a *fiesta de quince*. But they came.

It's funny how selective our memories are. Recently, when I asked my friend Mechas what she remembered of that night, she distinctly recalled not my *fiesta de quince*, but those childhood birthday parties with our evening dresses on in the daylight. When I pressed her about my turning fifteen, all she remembered was it being A *fiesta bailable*. A dancing event. But I remember *her* that night. I remember she came with her younger sister, who had just begun to walk again after a devastating traffic accident and had begged to come to my party, her first outing in more than six months. Mechas and I would become fast friends after that night.

Then there were the boys, pretending to feel cool in their suits, their hair slicked back, but really looking eager and respectful, emanating that sweet innocence which is forever lost later in the teen years. I remember Ika T., who was shy and gentle and who was regularly teased but still came and danced the night away and gave me the newest album by ELO, a rock band I never thought I would love. I remember the strains of the waltz, Strauss of course, and people coming to stand alongside the dance floor, their faces blurring into one as my father twirled me around and around and I tried to divine who would cut in first to be my next dance. I remember the sight of my older cousin, who spent the better part of the evening upstairs by the front door, anxiously waiting for a boyfriend that didn't arrive and staring futilely at the darkness outside. Never, I vowed, would I waste my time or joy waiting on some man.

I'll never forget Elisa—snide, disdainful Elisa: She had a snit when her parents came to pick her up. "But I want to stay longer!" I heard her whine. "It's not even one in the morning and we're all having such a great time!" As for me, I danced with my elbows down the entire night. I didn't make out with Felipe and our relationship survived another paltry weekend after that.

Looking at the pictures now, I see a girl who was only beginning to understand what she could be. My curls were still short; I didn't know that just five years later I would love them and grow them out, allowing them to be disorganized and wild, as they ached to be back then. My mouth is turned up in a slight smile that is almost a smirk, though it was actually a sign of shyness at being the center of all this attention. And although my mother always insisted that I stand straight, my shoulders are still slightly slumped forward, a vestige of teenage reticence.

In my green dress, I didn't look rebellious and certainly not sexy. My skin was unblemished and, despite everything I thought back then, I wasn't that chubby, just growing out of my baby fat.

But the dress has stayed the same, unruffled, unwrinkled, and pristine. Sometimes, when I go home to Cali, I look at it, so lonely now in my empty closet, and I try to imagine again how it felt to wear it that night, to begin the evening in my dad's arms, his hand placed firmly against my waist as he led me through my waltz, and to end it in Felipe's tentative ones, the ones that never dared climb up my back, the ones I never had to put up my elbows to restrain.

I look at the dress now and I have to smile. It's not glamorous, not strappy or gauzy or sexy. It's just a long green dress, the kind a little girl could wear to a ball.

Uprooted

BY Nanette
Guadiano-Campos

Tía Chelo made her grand entrance in a leopard-print spandex jumpsuit so tight, you could see every nook and cranny made by every tortilla she'd ever eaten in her seventy-one years on this planet. On the day of my quinceañera, she walked into the church right behind me in black patent-leather stilettos and fire-red lipstick, smelling of Jean Naté and Wrigley's Spearmint gum, oblivious to the "*Ay, Dios mío*'s" of the congregation and the mortified expression on Father Smith's Irish-Catholic face.

I knew something was wrong when I saw everyone staring at something behind me, rather than right at me. Tía Chelo's strong smells were almost enough to make me turn around, but the expression on my mother's face and her year-long obsession with this precise moment kept the smile glued to my lips and my eyes on the mark: Jesus staring down in pity from a life-size cross.

"Keep walking," my cousin (and escort) Juan whispered through a frozen smile. He had sensed my hesitation, felt the nails on my right hand squeeze his left arm. In my peripheral vision, I saw my father walk beside me, then behind, and then reappear with my *tía* on his arm. He led her to a pew where she actually waved at the crowd and popped her gum before winking at me in approval.

I had begged my mother not to throw me a quinceañera party. Just like three years earlier when I had begged her not to tell anyone in my family that I'd started my period. She had promised not to tell, so her betrayal had been worse than a physical punch to my throbbing ovaries.

I was leaving the bathroom with my evidence of womanhood discreetly wrapped in toilet paper, when the shushing and subsequent uncomfortable silence drew my eyes to my mother's ill-repressed guilt and my *tía*'s *pobrecita* gaze. My *abuelita*'s only response was, "Make sure you wrap it in newspaper and throw it in the outside *basura*."

I learned then that my mother was never one to keep things low key, and she was even more determined to tell the world that I was turning fifteen. After all, I was the first granddaughter on both sides of the family and it seemed that everyone suddenly felt a strong tie to their south-of-the-border roots. My impending coming-of-age was like the sweet, humid breeze of those southerly winds that reminded them of Mexico, where they began, long before I was even a thought.

The year I turned fifteen, my family moved to Uvalde, Texas, just as Iraq was invading Kuwait. It was 1990, the United States was threatening war with Iraq and George Michael had just released his long-awaited song "Freedom '90." It was right around this time that I had gone to my first New Kids on the Block concert and fell in love with one of the singers, Joey McIntyre, just as I was falling for Jody Ansley, a boy from my new neighborhood.

There were a lot of things going on, and the last thing I wanted was a quince. Please don't misunderstand. It's not that I didn't want to be the center of attention for a full day. Believe me, when you're the oldest of five girls, you don't get much of that. But I was fourteen. That age alone is enough to explain my hesitation at being the center of attention, but having just moved made it all the worse.

While I was trying to regain some semblance of normalcy, my mother wanted to relive her quinceañera through mine, color palette and all. My mother's mother wanted to show all of her church-lady friends that her eldest daughter was "doing pretty good," while my father's mother wanted to give me the party she'd never had. Tío Daniel, ever the comedian, volunteered to emcee. Tía Lupe, with her cushy new government job, wanted to sponsor the cake (she knew somebody in pastries). Tío Fernando had a camera, so he was the designated photographer, and my *abuelitos*, well, they fought over who'd pay for the food.

Meanwhile, I was starting the ninth grade at Ulvade's old elementary school campus because there was no room for the freshman class at the main campus. As a result, I had only about one hundred fellow classmates, half of whom should have been seniors.

Compared to San Antonio, Uvalde was a completely different world. I went to school with a bunch of cowboys and all my new friends were white. They spoke with southern drawls, chewed to-bacco, and swore on a Bible whenever the occasion called for it. They'd never been in the presence of a Tía Chelo or seen a giant crucifix before.

In keeping with the status quo, I told my mother I wanted a sweet sixteen instead, because that's what all my new friends had. They got to wear simple white dresses, a little makeup with foun-dation, rather than just lip gloss like good Mexican-American girls. I don't know about other Latino families, but mine believed (quite falsely) that only people with bad skin and prostitutes wore foundation.

Borrowing my friend Cristina's blue eyeliner was also out of the question. It wasn't fair. My girlfriends all got a special trip to Merle Norman to get a makeover before their sweet six-teens. I couldn't even put on a little powder to take the shine off. They got to dance in a banquet hall without having to go to church beforehand and be warned about the threat of hell after premarital sex.

If I had known my request would actually cause a rift in my relationship with my mother, who actually cried when I told her this, I might have thought twice before asking. My father was so upset at the fact that I'd caused my mother pain; he announced that it was a quinceañera or nothing.

"*¿Qué quieres, una* sweet sixteen? I knew we shouldn't have moved here, Carlos," my mom wept to my father. "She's angry at us, and this is how she gets back at us." She spoke as if I weren't in the room. "How can you even think of having a party at a hotel? A quinceañera is a public declaration of thanksgiving to God and your *family* for all that you have and a testament to what you have become. You cannot deny God what you owe Him."

She said it with such conviction that I felt like Judas Iscariot, betraying my family for a few gold coffers and a simple ceremony. The very suggestion of taking this sacred rite of passage and demeaning it by turning it into a "party" was committing sacrilege in her eyes and in the eyes of God Himself. Honestly, I didn't see what the big deal was. Fifteen, sixteen, a party's a party, right? Wrong.

You see, in my third-generation Mexican-American view of the big picture, it *was* just a party. One where I got to wear heels and dance with boys. But to my mother, my *abuelas*, and my *tías*, it was a sacred ritual. It was a pledge before God and la Virgen that I was still pure (which I was: The closest I had come to understanding the changes going through my body was that funny feeling I got between my legs when I heard George Michael's "Father Figure" or kissed Joey McIntyre's photo). Somehow, the party became the symbol of my womanhood. I'd been converted into a living homage to an entire sex. No pressure, right?

Part of that sacred right of passage was the time-honored tradition of the dress. White: a testament to the congregation of my virginity. Oh, God. It's hard even now to deliberately remember this part. I've tried blocking it out a thousand times. But there I was, with my mother and my *abuela*, trying on the puffiest of dresses, looking like a life-size cake topper.

"*Déjame ver,*" Abuelita would say, putting on her Elvis-style tinted bifocals.

"Turn around, *hija*, lemme see your backside."

"*¿Qué piensas?*" my mother would ask.

"No, no. It looks like a wedding dress," I said.

"*Otra*, Nettie, *pone la otra*," said Abuela.

"They all look like wedding dresses, Abuela," I said. "Why do I have to wear white?"

"Don't sass me," my grandmother said, gritting her teeth.

"Besides, what will people say if you don't wear white? Your *cola* is already too womanly as it is."

Seeing that I was close to losing it, my mother added calmly, "What your grandmother is trying to say is that white symbolizes purity, *mija*. It's a declaration to the world that you are a virgin."

I found this incredibly hypocritical. Everyone in my family knew that by the time my cousin Mona had her quinceañera, she had already broken in her chastity belt several times over. And she wore white.

I tried on dresses for two days. I cried for two nights. I went to my father and begged him, pleaded with him for my right to self-respect. I played on his regret at uprooting my four sisters and me, sending me to a new high school in the middle of a tumultuous freshman year.

"Daddy, they will laugh at me," I begged. "I can't show up in a puffy dress. Nobody here dresses like that. Please, Dad. Please tell her." We both knew who "her" was, and we both knew what I was doing was such an act of betrayal, going behind her back to get what I wanted. I was on dangerous ground, but desperate times call for dangerous measures, and I could feel my father wavering. I decided to follow my mother's lead and turn on a few tears for effect.

"I'll talk to her," he said gruffly. I knew I had him; my father couldn't stand to see any of his girls cry.

The issue of the dress was settled. I don't know what my father told my mother, but the next weekend, we went shopping without my *abuela*. But it turned out to be even more unpleasant than the other times. My mother was so cold, I could have literally scraped ice from her shoulders. I had gone to my father, had sidestepped her authority, and that was beyond mutiny, beyond forgiveness.

The day went from bad to worse, and my initial joy at winning round one was rapidly replaced by the tension every Hispanic daughter feels when she's disappointed her mother. For the sake of peace, I agreed on a lace number, tight on the body and flared out at the knees, what my mother called a "mermaid dress." It still had puffy chiffon sleeves and a big bow on the ass. (She had to throw puffy in somewhere.) The bow was an attempt to cover my growing backside. Personally, I think the bow screamed, "Look at me! Aren't I huge?"

I had always had a big butt, even as a child. I wore larger-sized underwear to avoid constant wedgies. But when puberty hit, it seemed to grow at full speed. I began to feel uncomfortable in my own body—my ass a constant reminder at how un-white I was. I could never wear the Wranglers all the cowgirls wore because I could never get them past my knees. I had to settle for leggings, "*algo con* give," my *abuelita* would say, nodding as I grabbed two pairs during back-to-school shopping.

After the dress fiasco, we began to plan for the actual event. I have to say I started to become more excited about it once I had the dress. My mother pulled out her own quinceañera pictures. She had them stored in a cedar chest with her wedding gown, love letters, and a collection of old magazines. As I looked at her pictures, I marveled at how young my mother looked and how pretty she was, at how black my grandfather's hair was, and how much my *abuela* still looked the same.

I thought, okay, this might not be so bad.

I was wrong. I couldn't believe the change in my household once the planning got under way. My *abuelitas* drove over from San Antonio almost every weekend to glue little pearls onto the lace of my mermaid dress. And to my surprise, my *tías* came over to plan the entertainment end of the party. This worried me, since

I had invited so many of my new friends and my reputation was at stake.

I would roll my eyes and complain about the music, about the *perlas*, about the colors for the *damas'* dresses (dark purple and fuchsia). My *tías* would glare at me and shake their heads, patting my mother and saying things like, "*Mira la desgraciada. If only our mother had given us all of this.*" To which I always wanted to reply, "She did. I've seen the pictures." But my staunchly respectable upbringing was deep-rooted, much like my dislike for Tejano music. So I kept my mouth shut.

My father, on the other hand, has always hated Tejano music, polkas, and anything else "oompah." His father had reared him on opera, the music of his paternal ancestors, and he listened, with pure rapture, to the voices of the Three Tenors every night before bed. I thought that if anyone would understand my angst at the proposed musical repertoire, it would be my daddy. I felt another stealth attack coming on.

"Please, don't let them play that music. What will all my friends think?" I hit a wrong chord with my father, who suddenly felt a strong tie to his Mexican roots. His adamant opposition to peer pressure topics always led to the question "If they jumped off a bridge, would you?" This would cause many eye-roll-induced headaches throughout my adolescence.

"They're not your friends if they don't accept you for who you are."

"They do accept me for who I am. I just hate Tejano music."

I could see my father's fear at the potential damage of caving into all my adolescent whims. "Well, *mija*, you are just going to have to suffer through whatever music the DJ decides to play. After all, you aren't paying for it."

Round two, I lose.

I sulked for weeks, but it only seemed to encourage my mother, whose winning the second round was like a shot of B-6 to her ass. She began a Mexican music marathon at home and in the car. I knew she was doing it to get on my nerves, especially since she gravitated toward Barbra Streisand, Barry Manilow, and anything Motown.

"Where did you get that music?"

"From your *tía*. She says it's all the rage."

She said it with a smirk, pleased with my obvious disgust, and proceeded to cumbia with the vacuum. Okay, I thought. Two can play that game.

I began to bring my girlfriends home after school to call her bluff. "Teach them to cumbia," I would tease, and she would glare daggers into me and switch her tape to *The Stylistics* or *The Best of Barry Manilow* (by the way, equally disturbing). My friends Cristina and Cara, who were delighted when I told them their names were Spanish, were actually intrigued by my mother. They loved coming over to my house with the blue den and the pink living room. They loved the green shag carpet and the smell of *frijoles* cooking on the stove in the afternoons. They stayed for *fideo* and beans, and I actually got my mother to teach them to cumbia so they wouldn't feel foolish at my party.

I couldn't believe it. Everything that had embarrassed me enthralled them. They bought Selena's cassette and practiced the cumbia daily to "*Amor Prohibido*," trying unsuccessfully to shake those hips to "Bidi Bidi Bom Bom." They developed a massive crush on my dad, who looked quote, "totally like a mobster."

"You're so lucky, Nanette," Cristina told me one afternoon.

"Why?" I asked, with my signature fourteen-year-old sarcasm.

"Your mother hugs you. God, she even tucks you in at night."

I didn't know what to say. I mean, it was embarrassing enough the first time Cristina spent the night and my mother had walked in with her *agua santa*, muttering in Spanish and making the sign of the cross on my forehead. I had wanted to die, to crawl under my bed and literally die from embarrassment. Of course I had never brought it up to my mother. Every Mexican kid knows some things are just too sacred. But I couldn't believe it when Cristina told me she envied me.

"My mother doesn't even believe in God," said Cristina. "My grandma died of an aneurysm when I was six. I saw it happen. Her eyes just sort of rolled to the back of her head and she fell asleep. It was awful. My mother said that a god that would let her daughter see something like that isn't worth believing in. Not that we went to church all that often before."

"Oh," I said, equally surprised by her story as by her declaration. The next time she spent the night, she asked my mother to bless her, too.

It was strange, my new life. My friends were book-reading, horse-wrangling, molasses-pouring Anglos. I felt sort of split down the middle, being one way with them, then going home to my family, where we said grace in Spanish and painted our walls in primary colors.

I had to adjust to a world where blond hair, blue eyes, and flat asses were the marks of true beauty and where I was suddenly sort of fat, at 5'6" and 120 pounds. On top of this, I was hairy, with eyes the color of Hershey's dark chocolate, black lashes, and leg hair, which I wasn't allowed to shave just yet. I had a monobrow, a flat chest, and an ass shaped like an upside down heart. I hated country music and two-stepping. The only place to shop for clothes was Wal-Mart, and the only place to go on a Friday night was a one-room movie theater that played movies I had seen months before back home in San Antonio.

To say it was culture shock for me is a massive understatement.

And to add to all the changes, I had to reconcile my "Latina-ness" with my new surroundings because I didn't know I could be exactly who I was and still have people like me.

And my insecurity showed. It showed in the way my father would look at me as if he didn't recognize me every time I complained about something. It showed by the way my sisters began to call me "white" to antagonize. It showed in the way I began to talk back to my mother, and in the way I began to look down on all things brown.

I was suddenly ashamed of my family, of my culture. Of the way my grandmother popped her gum every five seconds and fried her hair with countless perms. As a result, she had been left with little bald spots all over her head. I was embarrassed by the way my *tías* said "shurch" instead of "church," and the way they felt they had to bathe in cheap perfume to attract the "mens."

I never felt that shame more strongly than the day of my quinceañera, walking down the aisle in a used dress that I was convinced had been someone's wedding gown. From my hat that resembled an upside down salad bowl with pearls and tulle attached down to those tight-fitting Payless shoes dyed a pee-colored white to match my dress, I was simply a ball of nerves.

Standing in front of the church, I searched nervously for my friends, while my *damas ooh*ed and *ahh*ed over my dress. My *damas* were a mixture of cousins and old friends from San Antonio. My friend Geneva wore the dark purple *dama* dress with her girdle lines showing beneath the fabric. My dear friend Gwen had to have her dress specially made because of her weight issues. And since a local seamstress sewed all the dresses, she'd had to pay double and ended up looking like Barney, the TV dinosaur.

All the *damas* wore hats like mine, died to match their purple dresses that were worn with hot pink bows at their collars. Ten girls in ten drastically different sizes and shapes stood around the entrance to the church looking like a bouquet of dyed carnations. Somehow, in the light of the *iglesia*, the purple was much more, well, *purple*, and the hot pink was almost fluorescent. They were all Mexican, and therefore were not wearing any makeup (with the exception of Tía Valeria, who was a whole year older). It just seemed wrong that we were all dressed like grownups, but still looked like children. I was lucky, though. My mother actually let me wear mascara and lipstick, not lip gloss. That, in itself, was something to be grateful for.

I had let my *damas* choose their own escorts, because, well, because I just didn't care. So I didn't know any of the guys who stood in as *chambelanes*. One guy sticks out like a sore thumb in all the pictures, because he was so tall and so Anglo. To this day, I have no idea who he was.

The only boy in a tux I was well acquainted with was my cousin Juan, my escort and the bane of my existence since I was two years old. He was the older brother I never wanted. I was mortified that my father had suggested he escort me. I had begged my mother to let me walk in alone, rather than stand with Juan. But my requests went unheeded, and as I waited for the signal to start walking in, he stood next to me and let out the nastiest Big Red burp.

My Tío Fernando, volunteer photographer, signaled that it was time to start walking in. He was moving quickly, snapping shots of us like a pro. His camera had gotten so up in my face that for a while all I could see as I was walking in were big blue spots dancing in front of me. Once I recovered from my momentary blindness, I took another long, spotty look at my *damas* walking in with their ridiculously purple dresses. *Qué lastima*, I thought, sounding like my grandmother.

I tried to ignore the shame, along with the overwhelming smell of Jean Naté, White Shoulders, and my sister's Electric Youth by

Debbie Gibson. This interspersed with the scent of incense and my nasty cousin's body odor, which was starting to riot against his starched white shirt. That's when I caught a glimpse of Cristina's face in the crowd, the only white one on the right side of the church (most of my other friends were only coming to the reception, thank God), and the expression on her face. It reminded me of the story she'd told me about her grandmother's aneurysm.

I remember looking up at the giant crucifix leaning in toward me, then at Father Smith's impatient expression, then at my cousin Juan's giant size-thirteen feet—I wanted to pull up my poufy train and run. But where would I go? We were on the West Side, in the barrio, on Zarzamora Street. My only choices were the San Fernando Courts across the street, Handy Andy Grocery, or the Plasma Donation Center. I took a deep breath and prayed that afterward, my life would go back to normal, whatever that was.

Then Father Smith began the mass in his Irish monotone. "In the name of the Father, the Son, and the Holy Spirit." I went through the motions: Sit, stand, kneel. Sit, stand, kneel. I tried to focus on the seriousness of the day, but all I could think about was how tight my panty girdle felt (my *abuela* bought it for me, so my *cola* wouldn't look so big), how my cousin's pits were really starting to smell, and how I must have looked ridiculous in my second-hand dress and funny hat.

That's when I really embarrassed myself.

"Nanette . . . Nanette?"

Father Smith had been trying to get my attention and everyone had started to giggle.

"Yes?" My face felt like I'd eaten a handful of Red Hots.

"Do you want to say a few words to the congregation?"

Oh, God. Please make me invisible. I haven't asked for much. I've been a good kid, haven't I? I nodded my head tentatively and stood up with the help of my onion-and-wood-scented cousin. I could see my mother with her fancy handkerchief, ready to dab at tears that had been forming since the day she knew she'd conceived a girl fifteen years and nine months ago. I saw the proud look on my father's face, the irritation on my *abuelo*'s face (he was like the devil, uncomfortable in churches), and the smirk on my sister's face. I looked toward the other side of the church and spotted Tía Chelo and I wanted to crawl under the pulpit and die. She winked at me again, and I think I may have heard her gum pop, or maybe it was my *abuela*'s.

"I'd like to thank, *eeeeeeeeeeeeeeeecceeeeeeee*!" The feedback on the mike made everyone grab their ears, and turned mine redder than they already were. My mother's face tightened. This wasn't rehearsed. "Sorry. Uh, I'd like to thank my parents." Oh, shit. Sorry, Jesus. I forgot my speech. The one I practiced a million times, the one that said how special and giving my family had been, how supportive and sacrificing my parents were my whole life. The one that screamed, "We are a perfect family."

My eyes begged my mother's forgiveness. She was already dabbing, but I knew, unlike the rest of the congregation, that they were *not* tears of happiness. "Uh, this is a really special day. I'd like to thank my family for, uh, everything. And especially, uh, my mother and father for, uh, making me?" My daddy pulled at his collar. "I promise to the Virgin Mary and Jesus Christ, the Son of God, and to the Holy Spirit, and the Holy Catholic Church, and all the angels and saints, uh, that I will remain pure until my wedding day. Thank you."

If anybody, besides my mother and father, noticed how ridiculous my speech was, they didn't intimate anything. I think people were just ready to get on with the reception. My grandfather, especially, looked like he was ready to spontaneously combust.

When it was over, my uncle started in again taking photos of the Court of Honor, while people began making their way across the gravelly parking lot to the hall across the street. The DJ was already setting up, the caterer (a distant cousin), was getting the food ready, and my grandfather was undressing and asking for the keg. My grandmother was screaming, "Put that tie back on, Pach!"

I was standing in front of the little shrine to Our Lady when Cristina walked toward me, looking like one of the statues of the angels at the altar, and I just wanted to cry. I wanted to tell her it was all a big joke, a ridiculous gag, these people weren't *really* my family. (Although no one would believe that if they saw me next to my cousin Juan.) But she shocked me by smiling and hugging me so fiercely that I couldn't breathe.

"That was so beautiful," she said. She either hadn't noticed the fiasco, or she had slept through it. Cristina stood by and watched as Tío Fernando took pictures of me in front of the shrine of Our Lady with various people I was supposedly related to. While I was posing I couldn't help but notice the way Cristina was staring at my cousin Juan. And my nasty, smelly, snobby cousin was actually staring back.

We walked across the parking lot to the hall where my mother's parents were still fighting. Though I know my mother had worked really hard setting up the hall, my first impression was that my quinceañera looked more like a five-year-old's birthday party than a soon-to-be woman's. It was strange. I mean, here I was, celebrating the cusp of my womanhood, with all the pomp and circumstance of the day, and balloons were greeting me.

I had Cristina next to me and I was trying to feel proud of the hand-blown balloons of purple and fuchsia scotch-taped to the walls around the hall. They had already lost some of their roundness, since they had been taped up the night before. They just sort of hung there like deflated scrotums (I had seen pictures). The hall itself was in major need of repairs; the ceiling leaked in places and it smelled of mildew. Underneath that harsh fluorescent lighting, I noticed that they had pushed the big bingo setup into a corner to make room for the DJ table.

What can I say? We came from a poor parish. That explained the *cucaracha* that scuttled over Cristina's foot as she stood there greeting various family members buzzing around her like flies on a carcass.

"*Mira, la amiga de Nanettie. Es de allá. De Uvalde. Es anglo.*"

"That means *white*," someone whispered unsuccessfully.

"*Ay, sí. Qué chingada, ¿no?*"

Thank God the only Spanish Cristina knew was somewhere along the lines of "bidi bidi bom bom." As we walked away toward my corner table, set up complete with purple and pink fabric flowers, my family was still talking loud enough for both of us to hear them.

Most of my friends from school had shown up and we were all huddled in my corner, laughing and talking. Every time someone walked in, I'd look up to see if it was Jody, my dream crush. Though he had a girlfriend, I made sure not to invite her. My friend Cara said that he might show up. That's when Juan suddenly interrupted my thoughts of Jody's blond hair and his dark brown puppy dog eyes.

"Oh, my God," I heard him say. He started laughing and we all looked up to see what he was looking at.

"Wow, that's bright," Cristina said in her fake-nice voice. Several of my friends were giggling. I felt the panic start again in the middle of my chest, spreading its tentacles to the ends of my fingers and toes. Even my hair felt it in the tips of its fried hot-rollered follicles.

I scanned the room for my mother. This couldn't be the surprise they had all been planning for weeks. I spotted my mother and her face mirrored mine, only angrier. She marched over to my *tía*, the one who "knew somebody in pastries," who looked as mortified as we did. She shrugged at my mother, her face as hot pink as the cake.

"Excuse me, guys," I said to my friends, and I walked over to my mother. I could feel the tears forming in my eyes.

"It's hideous," my mother hissed.

"I don't know why she did this," my aunt was saying.

"Mom?" I asked, touching her arm.

"I know," she said, holding up her hand.

My grandmother walked up to us.

"*¿Qué es eso?*" she asked, pointing.

We all looked at it. It was a huge sheet cake, one layer. I'd never known any debutante to have a one-layer cake. It was the deepest fuchsia I'd ever seen. I don't think any of us knew frosting could be that color. And it was outlined in purple. Deep purple. The same purple from the neon sign outside the XXX video store off of Loop 410.

But the worst part, *definitely* the worst part, was the huge doll standing in the middle of the cake. She had a plastic head and was wearing a pee-colored frosted dress, just like mine. It looked unnatural and inedible. In fact, I would never taste my own quinceañera cake. More than half of it would be left over. And, as happens at many quinceañeras (at least in my family), before the end of the night, someone would steal the topper.

Just when I knew the tears were definitely going to fall, Tía Chelo walked over to me and said, "*Mira, Nanettie, su torta, qué bonita. Ay, ay, ay.*" I had to laugh. Only Tía Chelo would find a cake like that pretty.

The dinner went without incident, thank God, and by the time the DJ started playing music, I had mellowed out and started to have fun. My friends hadn't cared as much as I thought they would. It seemed they viewed my elaborate dress and the bright colors of the cake as a curiosity, much the way I viewed their cowboy hats, boots, and impossibly painted-on jeans.

And as far as the music went, I heard no complaints. In fact, at the first cumbia, Cristina and Cara jumped up and grabbed me by the hands, pulling me to the dance floor. They had been practicing for weeks, and wanted to show me their stuff. When the "oom-pah" music came on, it gave us all a chance to sit and breathe be-tween our dance music, and the grownups a chance to show off their stuff, the women sparkling like big sequined dolls, shaking their "oompahs" in time with their mustached men.

After a while, my mother came over and announced that it was time for the toast. Juan was nowhere to be found and neither was Cristina. I had noticed them talking and dancing all night, but, like, who cared? I mean, if Cristina could find him attractive, even with the overwhelming smell of *cebolla* emanating from his pu-bescent, oversized body, then good for her, right?

My mother and I and a few other girls went searching for them. But my father finally found them behind the hall. They were mak-ing out on a stone bench in front of a statue of St. Stephen, to my *abuela*'s horror and my great shame. After all, I had invited "that *puta*."

"What kind of friends do you have?" my *abuela* spat as she talked. "Young lady, I hope you know that I will not tolerate you becoming a *puta*."

"Abuela," I whined, shocked.

"I mean it, Nanettita. I am going to tell your mother and your *tía* about this. And you send Juan to me after you make that toast."

I nodded and went inside. As we raised our glasses for the toast, I was blessed with a fresh wave of onion smell from Juan's raised arm. I leaned over and whispered, "Abuela wants to talk to you." His face reddened and I smiled into the camera, knowing how much he feared my grandmother, remembering the time I ratted on him for watching a porno at her house one Christmas.

"Say it, don't spray it, putz," was his comeback.

"Why don't you do us all a favor and put on some deodorant?" was mine.

After the toast, it was time for the blessing. Tío Daniel started by announcing my mother's parents. They walked in, my grandfather self-consciously adjusting his tie, walking like his pants were too tight. He smiled his signature, constipated smile, the one that said, "My underwear is one size too small." The one that appears in every photograph I've seen of him from fifteen to seventy-five years of age. The only difference? His once jet-black, now baby blue hair, and a gold tooth for his upper canine. They stood before me as I knelt, my head down like a real-life princess, and they prayed, and then waved to the cameras.

Then, it was time for my father's parents to bless me. It seemed like it was taking forever, so I looked up from my saintly position to see my grandparents each holding the elbow of my great-grandmother Viviana. My great-grandmother, God rest her soul, was walking toward me at a negative five-mile-an-hour pace, all seventy-five pounds and four feet eleven inches of her, flanked on either side by my six-foot-tall grandfather and my five-foot-ten-inch grandmother. She shook with Parkinson's disease and kept asking my grandfather in Spanish where the bathroom was. The entire hall was still. But when she grabbed the crown with her shaking hands and lowered it to my head and began to pray, I was suddenly three years old again. There I was, grabbing Nilla Wafers out of her carefully kept cookie box, smelling her rosewater perfume, and listening to her speaking words in Spanish like music to my daddy. It was a beautiful moment, and one of the last ones I'd share with her before her death when she reached one hundred.

After the blessing, it was time for my big surprise. I had waited weeks for this moment and I was starting to wonder if it could possibly be more embarrassing than the giant bow on my ass.

"Sit down, Netty," my mother said, her face beaming in anticipation. They put a folding chair in the middle of the dance floor for me to sit on and dimmed the lights. I sat down, and the New Kids on the Block's "Happy Birthday to You" began to blare from the speakers. Five men, dressed in black pants and black T-shirts stood with their backs to me first and then turned around. It was Joey, Jordan, Jonathan, Donny, and Danny!! Actually, it was Joseph, David, Abel, Fabian, and Gabriel wearing masks with the New Kids' faces.

They began snapping in unison and, step by step (all choreographed by my mother), they re-created the dance the New Kids did in their video. I actually laughed until I cried. And they were good tears. I hugged my mother, and I think, for the briefest moment, she was genuinely happy.

It suddenly wasn't about pleasing the family or impressing her friends. I could tell by the way she squeezed me after the performance that it was about my genuine surprise. I felt like I was a little girl again on Christmas morning. My mother had given me what I loved because she knew me. And it sort of made up for the dress and the stupid hat. And for the balloons and that purple and fuchsia theme she came up with.

When my uncle announced it was time to serve the cake, I snuck out with my friends for some much-needed air. (Juan was also seriously reeking.) And though my heartthrob Jody never showed up, I still had Cristina, Cara, and the New Kids on the Block.

We sat and talked on the same stone bench that Cristina and Juan had made out on only two hours before. I watched my friends laughing over nothing and closed my eyes. I was swimming in the music of it all: the DJ's Motown tunes, my friends' laughter, a train passing by, a dog barking, a baby crying in the projects across the street, the chain-link fence rattling in the crisp November breeze. When I opened my eyes I saw Halloween lights over a door and an old woman with no teeth holding her purse tightly to her chest. She was waiting for the bus to pick her up. I glanced up at the sky and watched the clouds move across the dark blue like an army of horses and felt, for just a moment, such a sense of peace. I was right where I needed to be at that precise moment.

"Do you smell onions?" Cara asked, sniffing the air. "I smell pine," Cristina said. They were both right. Juan approached us with what I can only assume was supposed to be an alluring gaze pointed right at Cristina. Heeding my advice during the toast, he had apparently sprayed some kind of cologne or room spray to cover up his hideous man-smell. It didn't seem to bother Cristina, but it was giving me a headache, so I got up and went inside with Cara, leaving my cousin with my *"puta"* friend to pick up wherever they had left off.

It was time to fulfill my obligation of talking to all the strangers my parents had invited. I tried smiling through my oily skin and the pain of my tight pumps, and nodded to whatever whomever I talked to was saying. It was bizarre to have people I didn't remember looking at me with such love and tenderness. There were so many people there who hadn't seen me since I was "this high."

"I remember you when you were born. You looked just like your daddy. I told my husband, didn't I? That you looked just like Carlitos?" The man nodded at his wife, looking as uncomfortable as I felt. I kept that smile going as I moved around the room, until finally I was back on the dance floor, kicking off my shoes and pumping up the jam.

I remember Juan rolling up his sleeves at one point and break dancing. He looked like a total moron, but Cristina liked it. They would actually write to each other for about a year after that night, so I guess she wasn't a total *puta*, *abuelita*. My grandfather put his tie back on and danced a slow song with my grandmother. Tía Chelo danced with her common-law husband to every single polka that was played until it was time to clear the hall, shaking her groove thing and popping that Wrigley's gum.

As I look back on all of it now, fifteen years later, it's still hard not to laugh. The issue of the dress would come back to haunt me years later as I shopped for a wedding gown. My mother stuck her nose in the air at whatever dress I selected, saying it was "too ballerina-ish, too "poufy." You see how easily she forgives and forgets?

Things changed a little after my quinceañera was over. My mother and I got along better, perhaps because the whole ordeal was over or maybe because we understood each other a little better. I told her about my crush on Jody, my undying love for Joseph McIntyre, and my fear of never fitting in. She told me about her first crush, about her undying love for Paul McCartney, about her fear of never pleasing her mother. My father seemed to notice that I was growing up too. Whenever any of my sisters annoyed me, he would tell them, "Listen to your sister!" I loved it.

Back at school, word of my quince got around. I sat next to a girl in biology class named Isabel who I'd never talked to before. She noticed my ring, the one my *tía* Laura gave me at the ceremony, and asked if it was from my quince. I told her it was. She showed me a beautiful emerald ring that she, too, had gotten at her quince. We exchanged photographs and stories one day after school while waiting for our parents to pick us up. She surprised me by grabbing my hand and touching my ring, "This is a symbol of what we are now: strong, Mexican women," she said. "This will remind you for the rest of your life." Her dad drove up right after, so I never had a chance to respond. It was the first time it had ever occurred to me to feel proud of what I was.

At thirty, I'm just now starting to figure things out. I'm just now becoming comfortable with my large ass and my bright walls, thanks to women like Selena and JLo, who have helped make being Latina a social positive. But in a lot of ways, I've broken away from the traditions I grew up with. I'm no longer Catholic. I don't cross myself or use Agua Santa on my daughters at night. I don't pray to saints. But I do have a statue of San Antonio and my grandmother's handmade glass beaded rosary in my adult home.

I don't plan on giving my daughters quinceañeras. Not that having one was a bad thing. And yet, part of me is a little nostalgic for the ritual of a quinceañera—something about the idea of presenting your child to the world in a ceremony of pomp and circumstance that's usually reserved for the elite.

I will leave it up to my children. Whatever they decide is fine with me. The point is they will have a choice.

I imagine I can take my girls to Le Marche for a collective birthday bash instead in honor of their womanhood. We'll share a toast to the past and to the future over a good local wine from the Italian countryside. I imagine my girls and me sitting outside, under a clear sky, marveling at the beauty of the landscape, feeling the pull of our roots, the sweet, humid, southerly air.

I will smile at the utter perfection of the moment and think of Tía Chelo, who was too poor to have had a quinceañera. Perhaps her walking down the aisle at mine was actually the fulfillment of an unconscious adolescent dream? Or maybe it was her way of flipping convention the finger, telling us all to chill out, and just strut our stuff, preferably in leopard print.

Under that Italian sky, I will wink at the stars, like she winked at me that day, under those fabulously long fake eyelashes.

To Tía Chelo, wherever you are: You go, girl.

The Party Crashers

Quince Crashers

BY Malín
Alegría-Ramírez

*T*here's nothing worse than arriving to a party on time. Let me rephrase that. There's nothing worse than arriving at a *Mexican* party on time. There's this little thing called Mexican time, where the host has between forty-five minutes to two extra hours of flex time before the party actually begins. According to the white floral invitation, Cruzita Mora's quinceañera started at five p.m.

It was 5:02 when we arrived to drop off our gift. Colorful *papel picado* hung across the stadium-like hall. Stacks of chairs lined the wall to the right. The workers, dressed in freshly laundered shirts and dark slacks, were busy setting up the tables. I secretly think my mother wanted to arrive early on purpose so she could arrange the floral centerpieces herself.

Although showing up on time is bad, arriving in casual clothes and sick with the flu is worse, much worse. No amount of my sister's makeup could mask my red-rimmed eyes and runny nose. Her skin was a few shades lighter than mine, so the makeup made me look like La Llorona instead. Dressed in my dirty Nikes, faded Levi's, and a wrinkled plum-colored V-neck sweater, I wanted to hide behind the swan ice sculpture by the empty buffet tables. Maybe no one would notice, I hoped, trying to ignore the looks from the other early birds next to us.

"Check out what she's wearing," my twenty-four-year-old sister said, gesturing with her chin to the dance floor. An older woman with linebacker arms was chasing after a giggling two-year-old. Her long purple Tahitian wrap was tied haphazardly under her fourth chin. "Bet you ten bucks that thing falls off before the end of the night."

I snickered in agreement, then glanced around the spacious, nearly deserted room. Herbst Pavilion at Fort Mason is a 30,000-square-foot white tent located in the Marina district of San Francisco near the Golden Gate Bridge. From the outside railing you can see beautiful seagulls gliding on the wind and expensive yachts sailing near Alcatraz.

It was still early and guests wandered into the party looking lost. The room was enormously grand, with gigantic strings of twenty-foot-long pink balloons hanging from the ceiling like upside down peeled bananas. There were two dance floors set up, six bar stations, and at least a hundred tables. Three *viejitos* with polished guitars circled the room singing "*Besame Mucho.*" The floodlights were on bright, revealing all my flaws, which was fine because there wasn't anyone cute to flirt with anyway. A child's high-pitched wail pierced the cold air. This was not where I wanted to be.

I wanted to be home, sick in bed. Rent a couple of romantic comedies, order some wonton soup, and call it a day. But my mother called me from the mechanic's; the minivan had broken down by the side of the road for the second time that month. "Just come in with me for a second," she'd insisted. "Don't be *mal educada!*"

Quinceañeras were a community obligation. It was the time when everyone came together to honor a girl's transition into womanhood. It was too special an occasion to miss on account of the flu. My mother expected me to sit still and wait (with a big smile plastered on my face) until the appropriate time, which was somewhere after the toast and father-daughter dance, to skip out. But by the look of things, it was sure to be a *looooong* wait.

I rested my chin on my left hand to give the right one a break. My sister, who had also been persuaded to go, huffed loudly beside me. But Xochitl Mar, aka Xoch, never left the house without looking immaculate. It was like she was expecting to be discovered, at any moment, by some model or acting agent. Her slicked-back hair, dark Gucci shades, and the white fur-collared coat over her slender shoulders made her look ultra-standoffish. I felt downright shabby next to her. Xoch glanced at her pink Razr cell phone with a pouting expression on her glossed lips. She was probably hoping for a text from anyone who could save her from this party.

My mother was at the opposite side of the table with my cousin Rafa. Rafael Ortiz was top-down the hottest guy at the party with his sexy Latin lover, bad boy looks. He could make your toes curl, with his thick juicy lips, dark dreamy eyes, and his hard, twenty-four-year-old muscular frame. The guy was a Stanford alumnus, a superb dancer, and always the life of the party. A girl in a tight disco dress and big teeth passed by our table, extra slow, for the fifth time. She smiled coyly at my cousin. There was only one problem: Rafa was hella gay. He'd steal your boyfriend without even trying.

I heard my mother let out a hearty laugh. She was talking with an old white man holding a professional camera in his hands. "Take a picture of us," she insisted, posing with puckered lips. My mother had that natural *india* look, with straight black hair, mysterious dark eyes, and a soft olive complexion. At fifty something she still did not have one gray hair on her head. People swore we were sisters.

* * *

She promised that we'd stop by the party for only a few minutes. Before I knew it, two hours had passed. The building's plastic tent flaps blew wildly, letting in bursts of cold ocean air. An icy breeze sent shivers down my spine. I rewiped my nose on the moist napkin crumpled in my hand and looked over my shoulder.

"Okay, I'm ready to go home," I breathed, as more guests filed in.

"But they haven't even cut the cake," my mother protested.

"It'll be a while until things get started. I don't even think la quinceañera is here."

My mother squinted at me with a pinched expression as if I were barking like a Chihuahua. She pretended not to understand me on purpose. Couldn't she see how miserable I was?

Just then a familiar face appeared out of the crowd. It was my girl Francisca "Paca" Gutierrez. Paca was my old MEChA (Movimiento Estudiantil Chicano de Aztlán) buddy from San Francisco State back in the late 1990s. Paca was dressed in a retro-'80s ensemble from head to toe: gold metallic lamé coat, a black leather dress with leopard print heels, and super-teased-out hair. I was about to crack a joke, when I noticed her panic-stricken face. Paca was jabbering about some kind of trouble.

"You did what?" I asked. Suddenly, a coughing fit seized me and warm Coke blew out from my nostrils. My mother gave me a heated look and mouthed the word *cochina*. Paca leaned in to my right ear. The strong spicy scent of her perfume overpowered my nose.

"I *said*,"—she dropped her voice, peering slowly over her shoulder—"I went to the wrong party. Did you know that Ambrosio Zepeda is also having a quinceañera today?"

"You talking about the owner of Salsa Caliente Taquería on Mission?"

Paca nodded, shaking the coiled bangs that framed her round face. I couldn't help but roll my eyes. Ambrosio Zepeda had opened his business two years ago across the street from Delfina Mora's. He was always competing with Delfina, the owner of Taquería Fina. It was so annoying to see his fat face on flyers all over the neighborhood. The man acted as if he invented the taco. Fina's restaurant opened in 1979. It was a Mission institution.

My family were loyal customers ever since I could say, "Beans and cheese please." Taquería Fina was where we all gathered for graduations, birthdays, and funerals. I would never think of entering Salsa Caliente. It would be total blasphemy, even though I heard Ambrosio makes a mean *taco de pescado*.

Paca tapped her pointy leopard-print heel on the floor nervously.

"Uh-huh, and when I walked in he waved at me like we were old friends."

"Noooo!" Xoch leaned across the table and gasped. Rafa glanced over at us.

"What did you do then?" I asked.

"I served myself a plate of tostadas and I looked for you guys. You wouldn't believe all the *camarón* and *carnitas* he had." She gestured with her hands high above her head. "Oh, my goodness, the food was so delicious I thought I was going to die. They had this *mole* . . ."

"Okay, *mujer*, so the food was good. I get that, but tell me what you did."

Paca sniffed and shot me an exasperated look. "Well, like I was saying, I was sitting there eating my food when I noticed that I didn't recognize anyone at the party. Then they announced la quinceañera and a pencil-thin girl with blond cornrows came out. I almost choked on the piece of tortilla in my mouth. It wasn't Cruzita Mora!"

"No way," Rafael said, moving his chair closer to us.

"And?" I gestured for her to hurry up. This was just like Paca, she was always getting herself into trouble. The woman lived for drama. There was the time, in college, when she swore her neighbors were secretly videotaping her naked from across the street. Or the time she asked me to confront a stalker and the guy turned out to be a bill collector.

"And . . . ," Paca whispered forcefully, "I got up to use the bathroom, where I took a look at the invitation. This place is so huge," she gestured toward all of Fort Mason. "It was a simple mistake, all the buildings look alike. I walked out and saw a pack of people walking this way so I just followed them here!"

"I don't believe it," I said, looking at Paca and then at my sister. "Two quinces on the exact same day at Fort Mason by two rival *taquerías*."

"I'm sure Ambrosio did it on purpose," my sister added, crossing her arms in front of her chest.

"Hey, are the guys any cuter over there?" Rafa winked. I laughed and looked around the pavilion. A group of older men with bowed legs, cowboy hats, and Virgen de Guadalupe silk shirts strutted past us heading toward the bar.

"The boys at the other party? They're cute if you like jail bait." Paca rolled her eyes at Rafa and took out a compact from her soft black leather purse. She dabbed the peach-colored powder on her round nose, then gazed out toward the bar at the men who were quietly sipping Coronas. She frowned at the slender watch on her fleshy wrist. It was typical for guys to wait until their bodies surged with alcoholic bravado before they asked a girl to dance. Then Paca turned back to me with a desperate look in her eyes. "But what am I going to do about the gift I left at Ambrosio's party?"

Just then a loud commotion coming from the entrance made all eyes turn. A girl with a high-pitched voice was cursing loudly about the balloons. It was Cruzita's grand entrance. She was in a white hoop gown decorated with bows, pearls, sequins, and a sparkling tulle overlay. With the haughty confidence of an only child, she stormed across the room in tears. Her *tías* trailed after her like a hive, fussing over her makeup and tousled tendrils, while she whacked at her gangly boyfriend with her silver scepter to get out of her way.

I smiled. Good old Cruzita would never change. But I sure had. Fifteen years ago, a quinceañera was the last thing I wanted. All the *cha-cha* pomp and *fu-fu* circumstance was *so* embarrassing. I was a rocker chick at the time with purple hair, combat boots, and a fat chip on my shoulder. There was no way that I would wear a puffy princess dress in public. However, now, glancing at the pretty *dama* outfits, decorations, and stacks of gifts, I'd had a change of heart. I wished my parents had forced me to have one!

Paca shook my arm. "*Muchacha*, you have to help me get it back."

"You're crazy," I said, pulling my arm away.

"Please."

"Just forget it, all right?"

"I can't."

"Why not? There are so many people here. Cruzita won't notice that you didn't bring a gift."

"Ambrosio will notice," Paca said under her breath. Her eyes glazed over for a moment. "You see, I made Cruzita something special. Remember how she loved the blanket I made for Felicita's bridal shower? Well . . . I . . . I embroidered her a pillow."

"That's sweet," I smiled, patting her hand lightly and trying *not* to roll my eyes. I didn't understand the whole pillow thing. It represented something sacred about womanhood, I guess. Paca would know. She came from a very traditional Mexican family.

"You don't understand," Paca stopped to make sure I was

listening. "I embroidered a picture of Cruzita on top of a pillow with the words, '#1 Quince Girl in the Mission.' "

"No, you didn't!" I pressed my lips together tightly, trying not to laugh.

Paca nodded slowly. She looked down at her tight skirt and picked at a loose hem. "I thought . . . I don't know. Maybe it's lame . . . but whatever." She sighed as she shrunk back into her chair with a hopeless expression. "But now, I'm sure it'll cause all kinds of trouble between them."

I glanced across the table at my mother. She was talking to Delfina Mora, the mother of la quinceañera. Fina was a matronly woman with short hair and an easy smile. At her restaurant, she always stood behind the counter with a hair net, loose-fitted T-shirts, and a chile-stained apron. Tonight, she wore an elegant cream-colored pantsuit with champagne pearl earrings and fresh-looking makeup. Despite the makeover, the dark bags under her eyes gave away her fatigue. Quince planning was no joke, and this party in particular took a year's worth of preparation. My heart swelled at the sight of her. I thought about all the years of sweat she slaved in the kitchen to provide her family with a top-rate education. Fina was a testament of strength, motherhood, and independence in our community. The idea of Paca's gift at Ambrosio's party made my skin crawl. Damnit, I thought, biting down hard on my lower lip. Why couldn't Paca just get something from the Target registry?

"Maybe," I hesitated, studying Paca's reaction, "we could just slip in . . ."

"And get the gift!"

My stomach tightened. "How about we just check it out?"

"We could totally get it. It'll be so easy!"

Before I could protest, Paca grabbed me into a tight embrace.

Her arms were strong, body-builder strong. She could crush my petite bones if she wanted.

I put my palms face out, trying to calm her. "I'm not making any promises. I just said let's have a look, all right?" I got up to leave quietly. There was no reason to call attention to this matter. But as my chair scraped the tile floor, it made a loud farting sound. The family at the next table turned, shocked. I smiled, feeling my cheeks grow hot.

"Um, let's go."

"Don't worry." Paca stood up and ran her French-tipped nails through her big hair. "No one will notice."

Could it be that easy?

"Where you guys going?" my mother asked with Fina at her side.

"We'll be right back," I said, hoping that she wouldn't ask too many questions.

"Yeah, right back." Rafa stood up and downed his rum and Coke. Xoch started to fix her hair in her small compact. Obviously they both planned to come along without a formal invitation.

I pulled them a few feet away from my mom.

"What do you think you're doing?"

"Oh, c'mon." Rafa flashed me one of his adorable pearly-white smiles. "I'm starving and . . ." He glanced over his shoulder at the empty buffet tables. "Paca needs our help."

I looked at Paca and then at my family. My instinct said it was a big mistake. But it was already too late to back out.

"Oh, all right." I shrugged. "Just do me a favor and stay on the down low."

Rafa gasped dramatically, putting his right hand to his chest, and pretending to hold back tears. "I'm offended you would even have to say that. After all we've been through. Who taught you how to shave those Chewbacca legs of yours? And when you got that nasty case of—"

"Enough!"

Rafa smiled triumphantly.

"You dork."

"Slut," he retorted.

"Hoochie."

"Skankzilla."

Paca grabbed our wrists like disobedient children. "*Ay*, you two," she griped. "We don't have time for games. The gift."

Rafa and I followed behind Paca and Xoch in a snakelike line toward the front entrance. A pack of unruly kids, twin strollers, and *viejitos* with metal walkers entering the *fiesta* forced us to fan out and find our own paths to the exit. A DJ came on and tested the speakers calling out "*Sí, no, sí, no*," over and over again on the ear-splitting PA system.

I spotted the EXIT sign and pushed my way past a tubby guy with a small head drenched in Old Spice cologne. The sun had set, but the night sky was lit with flashing yellow car lights waiting to park. A red monster truck with naked-woman flaps honked at a purple van parked by the door. A bunch of cute boys with long sideburns were busy unloading instruments. BANDA TERREMOTO was stitched onto the backs of their yellow satin shirts.

"Oohh, *papacitos*," Rafa exclaimed under his breath. Then he nudged me in the ribs. "I wonder if they're on my team or yours."

With family Rafa played the obedient model son role. But on the street and especially at the clubs he was a *tigre* and "out" on the prowl. As his older cousin, I was very protective of him and worried about his safety, especially here, in *macholandia*. An innocent flirtation could end in a bloody fight.

Rafa turned to the musician holding a large brass tuba.

"*¡Oye!* My cousin thinks you're cute. You married?"

"Rafa!" I cried, pulling him back by the crook of his arm.

My cousin turned to me and snapped, "*Ay, no te hagas*, you know he's hot."

"This is not about me," I tried to steady my voice and not look into the tuba guy's chiseled Alejandro Fernandez–like face. "We came to get the gift, not a date."

"Don't act all like that," he said in a loud voice. "I saw you checking out his *chile verde*." Rafa winked as he sashayed over to Paca and Xoch. They were both wiping tears from their eyes at our spectacle.

"Ha ha," I said sarcastically.

Ambrosio Zepeda was an opulent man with beady black eyes and a beak nose. He drove around the Mission in a gold-colored Hummer with a license plate that read: TACO KING. His thick dark hair was woven into straight cornrows and his fingers, ears, and hairy chest were always covered in bling.

According to Paca's report, Ambrosio had rented Hall C for the quinceañera reception. Was it just a coincidence that he had decided to have a party here, I wondered? I didn't even know he had a daughter. Maybe he didn't? Maybe he was getting into the quince production business now too? I was sure that Ambrosio was trying to one-up Fina. He was so competitive he even hired Barbie-built *morenitas* in *dos XX* tube dresses to stand in front of his establishment. Pathetic. His party *had* to be twice the size, with thousands of guests, and probably even the mayor. It would be easy to sneak in, I thought, as we headed down the dark empty alleyway from Hall A to Hall C.

"Do we have a plan?" Xoch asked walking behind me.

I turned, thinking quickly. "Let's scope out the scene first." I gave Rafa and Xoch my warning look. "We can't screw this up. Key word here is *invisible*. Got it?"

They nodded, but their eyes twinkled with mischief.

"Don't worry so much," Paca said, putting her hand on my shoulder. Her gesture was comforting. I took a deep breath to steady my nerves and sniff back my *mocos*. "Everything will be all right." Paca walked toward the Black Eyed Peas jam blasting from inside the door to Hall C. She started to sway to the beat. "I got a good feeling about this."

Paca's vibes were contagious. Anticipation replaced my cold symptoms. My heart started to pound with excited energy as I took my first step into the hall. Everything was white. The blinding floodlights made me double back to the doorway. Shading my eyes with my hand, I noticed a mere six long tables in the boxlike, stuffy room. The aroma of cooked *bistec* and lemony disinfectant clung to the thick red drapes like static cling. There were pink tissue paper bows tied to the back of each chair like presents and bunches of blue and white balloons floating like center bouquets at the middle of each table. The tables faced a small dance floor and DJ table.

"Oh, shit," Xoch mumbled from behind me. Oh shit was right. My heartbeat started to race, pounding loudly in my head. This was not what I expected. A young girl with braces and a black cocktail dress bumped into me without apologizing. I gestured to Xoch to close her mouth.

"There's nobody here," Rafa whispered.

"I see that," I said, a bit too loudly. The room was hella small. It was about one-tenth the size of Fina's *fiesta*. I could make out every pockmark on the young DJ's crater skin as he nodded to the beats he spun on the opposite side of the room. The guests were all under eighteen and dressed in their finest church clothes. They sat quietly at the tables waiting to be told what to do. What was Ambrosio trying to prove? I became even more suspicious.

Disgruntled guests tried to pass between us. We were blocking the buffet line at the back of the room. I grabbed a paper plate and served myself a heap of rice. I didn't want us to appear even more out of place. My street clothes and sniffling red nose said enough. Rafa and my sister followed. This was crazy, I thought, looking around the room. Paca was standing at the end of the food line, smiling brightly and dancing in place. Why was she so happy?

At the center of the dance floor, piled high like a Mayan temple were the quinceañera's gifts. I scooped some beans and *camarones a la diabla* onto my plate. Think, I told myself as I munched on a delicious piece of shrimp. We had to get to those gifts.

"Ess-cuse me," a thick Spanish accent said behind me. I turned to find the frown of a masculine-looking woman wrapped in a turquoise satin evening gown, a short head of curls, and a Frida Kahlo unibrow.

I smiled, my cheeks stuffed with food.

The woman fidgeted with the shawl on her shoulders as if uncomfortable in a dress. She stared disapprovingly at my jeans and sneakers. "Who are you?"

I sniffed loudly, eyes wide, looking for Paca. Then I cleared my throat. "*Hola,*" I offered my free hand. The woman shook it cautiously and then looked behind me at my sister and cousin, greedily eating from the buffet as if they'd never seen food before. "Oh, hi. We're friends of Ambrosio."

"Ambrosio invited you?"

I bit my lip, hoping that would fly.

"And your invitation?" she held out her hand, expecting me to hand it over.

I turned to my sister, who was breathing heavy on my neck and making my hairs stand on end. "It's in the car," she jumped in. "I can go get it."

The woman furrowed her brows as if trying to see into my head. She released her breath and nodded. "That would be wise."

Xoch pulled down the large-framed Gucci shades over her eyes and snuck off with her full plate. Praying that Xoch wasn't jumping ship, I grinned at the woman who had yet to introduce herself. I wished she'd move on, talk to other guests. But she stood right next to me, growing roots.

"Where's Ambrosio?" I asked with a nonchalant tone.

"He'll be right back," she said, grasping tightly to my wrist. Panic coursed through my veins. "I'm Bernarda. His sister. Don't go."

My head started to throb, but I couldn't give in to my aching body. "You're funny." I forced a laugh and tried to wiggle free. "Of course I wouldn't go. This has to be the best quinceañera I've been to in a long time. I've been waiting for this day ever since I got my invitation. Marked it in red on my calendar."

"Red?" Bernarda asked, shocked. "You a *norteña?*"

Was this woman a gang leader? Did I just make a major gangster faux pas? Was Ambrosio involved in the drug underworld? Damnit!

"I meant blue. Did I say red? I hate red. That color is for sissies and chumps. Can't stand that color."

Bernarda squinted her eyes so tight I swore they'd disappear behind her cheeks. She was scanning me. Where was Xoch? And Paca? My sweater felt tight, suffocating. This was madness. *Abort mission*, I screamed to myself. Our best bet was to cut our losses and sneak out now, quit before I shoved my foot even deeper into my mouth.

Abruptly, the lights went down. I blinked trying to make out the dark shapes moving around the space. Bernarda's heavy breaths clogged up my ears. Then the DJ's booming voice sliced through the room. "Here's a special request from the birthday girl to get the party started." A burst of multicolored hues lit up the profile of a tall lanky girl in a pink princess gown standing at the center of the dance floor. A thumping rap beat blasted from the speakers and bounced off my chest. The quinceañera in blond cornrows shook her shoulders and grooved to the rhythm of the electric slide. Three girls in matching peach prom dresses joined her. A high-pitched cheer erupted from one of the tables and a dozen teens jumped up to dance.

Out of the corner of my eye, I noticed Paca's fuzzy 'fro. She was creeping toward the table of presents right by the dancing quinceañera and her court. My heart started to pound wildly. I had to keep Ambrosio's sister and everyone else for that matter from noticing her.

Rafa must have also noticed, because he grabbed my hand and said, "Let's dance," before Bernarda could object. We squeezed in between two tables filled with hungry guests and headed toward the dance floor. The disco ball sparkled its twinkling lights across the room. I stood next to Rafa, following the easy sway of his hips. I glanced over my shoulder looking for Paca. Did she have the gift? But as I turned, I collided with Bernarda. She was dancing right behind me! The woman was sticking to me like a *piojo*.

Over Bernarda's shoulder, I saw a hand waving frantically from the farthest corner of the dance floor. It was Paca. She was leaning against a side exit. The door was propped opened to let in a cool breeze. It was a perfect escape route. A pink ribbon hung out of Paca's jacket. The present. I gestured with the tilt of my head to Rafa that it was time to go.

Bernarda must have seen my gesture, because she moved to block my path. "Where you guys going?" Rafa twirled and disappeared into the crowd.

"I'll be right back," I yelled over the deafening music. "I need some air." My heart thumped wildly as I pushed passed Bernarda. I could feel her eyes burning a hole into my back. I had to lose her,

I thought as I ducked back toward the buffet. I squeezed past a couple and scurried around the rim of the room like a mouse. Bernarda tried to follow me, but a net of moving bodies swept her back into the crowd. I hid behind a large man in the far corner who smelled of tobacco and leather.

The music changed and I peeked out around the man's beer belly. There was no sign of watchdog Bernarda or Rafa. Where was he? I caught sight of him standing by the front entrance. I breathed out with relief when he ducked out the door. I couldn't believe that we were actually doing this. A rush of excitement, fear, and anxiety washed over me.

Quickly, I pushed past guests leaning against the wall. Paca was still by the door when I got to her. She was standing as still as a statue. What was she doing here? Hello? "Move," I said between clenched teeth as I pushed her through the door.

Outside, the cold cigarette-infested air smacked against my face. I stood there trying to catch my breath. Three body-builder types with shaved heads and tattooed knuckles were joking among themselves and smacking a little guy around like he was a *piñata*. Run! my mind screamed. Daddy Yankee's chorus filled the air and I begged the rapper to give me enough *gasolina* to make it safely out of there. My feet moved, trying to put as much distance between me and Ambrosio's party. The sounds of laughter, Paca's clicking heels, and my pounding heartbeat enveloped my ears. Dark shadows loomed behind trashcans, empty doorways, and down the street. I imagined those shadows coming for us. Fina's party was less than thirty feet away, just around the corner. If we could just make it to the pavilion doors, I knew we'd be safe.

"Hey you!" a woman's voice sliced through the air.

I was about to turn back when Paca grabbed my hand and told me to keep moving.

"You, in the gold lamé jacket!"

"Don't stop," Paca commanded, walking quickly alongside me.

"Get those girls," someone yelled.

"Oh shit!" Rafa cried from up ahead. Paca took off flying like a comet with leopard-print heels. I wanted to tell them both to stop. That running would only make us look *more* guilty. But I had bigger problems. Heavy footsteps were gaining up behind me. I pushed my stride even farther, passing Paca.

"Aggghhh!" Paca screamed behind me.

It sounded like someone had grabbed her by the hair and pulled her to the floor. The sensible thing would have been to stop. Paca was my homegirl. She'd been there for me through my college adventures, listened to my sob stories, and held my hair when I threw up. Could I be that cruel to abandon her in her time of need? I decided to keep going. Someone had to be around to tell the tale. Besides, this was Paca's problem, not mine. She's the one who went to the wrong party.

The lights of Fina's *fiesta* burned bright in the middle of the block like a beacon before me. Safety, I thought, releasing my breath. Paca would get over it. She had to forgive me. I was giving her a ride home.

"Not so fast," a manly voice called out and grabbed me by the shoulders. I screamed out, hoping that someone would come to my rescue. Strong hands turned me around. It was Bernarda. Her nostrils were flaring like a wild bull. Beads of sweat trickled down her wide forehead. Damn, I thought. When I looked back, I saw that a big guy with a Pancho Villa mustache was holding Paca by the forearm.

"This is a huge misunderstanding," I said, putting up my arms defensively in case Bernarda felt like slugging me.

"Misunderstanding?" she said, making a sour expression. A

crowd of people from both parties swarmed around us. The on-lookers held their breath, hoping for a down and dirty catfight. Guys were making bets and shouting angry threats from around the circle. This was not part of my plan. Where was Rafa? I had to stop this before it got even more out of control.

"Look," I cried, pulling the gift from under Paca's coat. Bernarda's eyes were red with anger. She looked like she was about to stomp all over us like *cucarachas*. Words sprang out of my mouth like a fountain. "My friend went to the wrong party, okay? She made a mistake. We didn't want to make a big deal so we thought we'd just get the gift back. We didn't want to interrupt your party." Bernarda's deep frown and furrowed brows told me that she didn't believe me. "See," I said, opening the gift.

"Stop that!" Bernarda cried. The crowd leaned in to hear. Someone yelled, "That's not right." Another voice said, "You better stop her!"

I ignored the voices and ripped off the bow and metallic paper as fast as my fingers could move. Beads of sweat were stinging my eyes. My head started to throb. I felt feverish again. Maybe I'll faint and then they'll have to stop and rush me to emergency. But I didn't faint. I was still standing and my fingers were still moving, tearing at the gift. "Look, see, there," I shouted, pushing the stitched pillow in Bernarda's face. "That's a picture of Cruzita. Cruzita Mora." Clearly, the girl stitched onto the pillow was not the pale blond cornrowed quinceañera at Ambrosio's party. It looked kind of like E.T., but that's not the point. "You see, Cruzita's having a party here too." I gestured to Herbst Pavilion as if that would explain everything.

For one long beat, nobody moved. Paca stared at me wide-eyed. Bernarda stared at the pillow. *If this big woman comes at me*, I thought, *I'll go for the eyes*. I'd never been in a fight before, but it seemed like the best thing to do. Her blank expression unnerved me. I was dying for this to end. I wanted to go home and be sick in bed, where I belonged. The sound of my uneven breath made the seconds feel like hours.

"So, that's why you lied and ate our food," Bernarda spat on my face. *Ouch*, I thought, wiping my cheek.

"Kick her butt," some girl yelled out from the crowd.

"Yeah!" a guy shouted.

Slowly, Bernarda looked around her as if she'd just noticed the audience. "You should have said something."

"I'm sorry," I winced, hoping that the apology would be enough. Okay, so maybe we wrecked her party and caused a big scene, but this was a quinceañera.

Bernarda looked me straight in the eye, making me feel icky, like a piece of crud stuck in her teeth. She opened her mouth to say something, but then changed her mind as if I wasn't worth the breath. Bernarda turned without a word and walked away.

"Ah, c'mon," someone moaned. "Chickens!" another person cried. The annoyed crowd griped a little more, and then lost interest. A blast of trumpets coming from Cruzita's party announced Banda Terremoto. Couples hurried inside the pavilion to dance. Xoch ran over, huffing loudly.

"You guys okay?"

I stared back at her in shock. "Where were you?" I asked in an annoyed tone. "We almost got our butts kicked."

Xoch smiled and shrugged her shoulders. "Oh, c'mon, you were just fine. I saw the whole thing from behind the car. I was going to jump in, but you had it all under control."

"Yeah, right." I rested my hand on my chest to calm my beating heart. Was that really it? I looked around for some kind of trap or twenty more thugs waiting to jump me. But there was nothing. *Nada.* It was kind of anticlimactic. Oh well. I peered down the alleyway where Bernarda had disappeared.

It was dark and there was no evidence of her anywhere. I thought about what Bernarda had said. Maybe we should've told the truth from the start? Was this one of those moments when the heroine realizes that honesty is the key to happiness? Then Rafa walked over, smiling triumphantly. The mischievous look in his eye got me thinking. Sure, we could've gone over and explained the situation, but where was the fun in that?

Paca and the guy with the Pancho Villa mustache were picking up pieces of tissue off the floor. They were smiling at each other. I couldn't help but roll my eyes.

"*Oye*," I said, pulling Paca up by the wrist. "Go in there and give Cruzita the gift. Do it now," I insisted. "Before anything else happens."

La Mamasota

BY Adelina Anthony

*I*t was the summer before our sophomore year and quinceañeras were as common as those unwanted zits that kept appearing on our adolescent faces. But there was one quinceañera party where the allure of a woman's body left me questioning all of my straight-girl assumptions.

There I was, making my best friend's waterbed undulate as I shifted my weight side to side. "C'mon, Annette," I said,

"Nobody's gonna even notice." But Annette ignored me and kept on picking at the blackheads on her nose.

She had been standing in her black slip and black bra for over an hour. Her dark brown hair was cut and permed into a short bob. We both had our bangs perfectly trimmed, and with just the right amount of Aqua Net, we could puff them up into little hovering clouds above our foreheads.

All along I tried to hide the subtle fact that I actually enjoyed looking at Annette's svelte back; it had a peculiar way of arching near her backside like the curve of a spoon's handle. Being a good girl, I also tried not to linger too long on her well-defined calf muscles, her plump thighs, or her firm breasts in the mirror's reflection.

If she ever caught me, I could brush it off by saying I was just looking at her "patio design." That's what we called her peculiar skin disorder, an almost imperceptible skin discoloration throughout her body. If you looked real close her skin resembled a tessellated patio, the kind made up of smooth marble stones and similar shades.

I tried another tactic.

"We're gonna miss the beginning of the dance."

"Uh, Lina-lu, we *are* the beginning of the dance." She said, putting on an exasperated face.

She had a point. All summer long, we'd been crashing one quinceañera after another. And sure enough, our *sinvergüenza*-selves didn't need a guy to ask us out onto the dance floor. We had each other and our imitation Madonna and Cyndi Lauper moves that we learned from watching countless hours of MTV in her living room. We were adherents of the "girls just wanna have fun" motto.

But I really wanted to be on time for this quinceañera. My appetite was insatiable as a teenager and my stomach rumbled like an angry dog that night. Plus, *raza* does not pass up free food, especially a warm *plato* of tortillas, *arroz*, *frijoles*, enchiladas, *pollo*, and a juicy jalapeño on the side to boot.

Like the other quinceañeras, we hadn't "officially" been invited, but that's okay. We lived in Southside San Anto' with its mix of welfare, poor, and lower-middle-class *gente*. Everyone knew that St. Lawrence Church's dance hall was a community affair. Whether it was a wedding or a quinceañera (sometimes it was hard to tell the difference, the brides were so young), you just had to show up. Once you were there, you were bound to recognize someone from Sunday mass or from around the neighborhood.

Annette was taking forever. Probably because her one-and-only-I'm-going-to-love-you-forever-no-matter-what-guy, Ricky, was going to be there. And though Ricky never responded to Annette's love letters, he didn't ask her to stop writing them either. "That has to mean, something . . . right?" she'd ask me. I would nod and stay silent. The last time I tried to convince Annette she should just move on and get over him we almost had a falling out. So I decided to support her delusions no-matter-what-because-we're-best-friends-forever.

So when we heard that La D-D, Deidra, was going to have an official quinceañera, well, *nos morimos de la risa*! The thought of La D-D (we purposely stuttered our D's to emphasize her breast size) in a white dress and a tiara was too much to imagine. Asking La D-D to don a white dress was like sticking a pair of white angelic wings on *el diablo* himself. Obviously, whoever La D-D's mother was, she didn't know her *niña* had already secured her bad-girl reputation by dating most of the bad boys in school, including Ricky.

Ricky and La D-D had been dating steady since the end of our freshman year. She was one of the few freshman girls to attend

McCollum High School's Junior-Senior prom. When Annette heard that the love of her life, Ricky, had asked La D-D to go, she brushed it off saying, "Whatever." Ricky and La D-D became one of those inseparable *chicle* couples—stuck on each other as if they were already married. You would spot them making out telenovela-style wedged in between the gym wall and the auto shop building, safely *escondidos*. We all envied their outright passion.

Despite Ricky and La D-D's lock-mouthed sessions, Annette kept writing her letters of devotion, kept slipping them in the vent of his top locker, kept hoping for something.

Okay, so sometimes, I wanted to slap my best friend—but hey, who was I to judge her? I still felt my heart jump and my *chones* twitch every time I caught a glimpse of my third-grade crush, Roberto. But at least I didn't humiliate myself and religiously write him letters. Maybe one letter a year. Fine, two.

"Okay, Lina-lu, let's go see my honey in a tux," Annette declared finally. She slipped on her tight polka-dotted miniskirt and her black lace top.

"Your nose is red now," I told her.

She looked back into her mirror and dabbed more Almay on her nose. At that moment her mom knocked on the door. They shared a bedroom, the way most not-so-rich families do.

As the eldest of eight, I already knew privacy is a luxury. That's why I loved hanging out with Annette, the only girl child in her family and the baby. Though a little spoiled, she was really sweet when she wanted to be. We had been best friends since seventh grade, when we found ourselves struggling to finish a model of our Texas state capitol with straws, posterboard, and a liter of Big Red soda. (Don't ask . . . We didn't get a good grade.)

"Are you girls ready?" yelled her mom from the other side of the door. If it hadn't been for her mom's willingness to chauffeur us to and from everything, our social life would have been as boring as geometry class.

"How do I look now, Lina-lu?"

I approved with a nod. She was almost an exact replica of the Betty Boop character she so loved and plastered on her bedroom wall, school notebooks, and locker. Her face was all big eyes, long black lashes, and moist red lips. Too pretty for dumb Ricky. No matter how many times I told Annette she was one of the most beautiful and smartest girls I'd ever met, she needed to hear it from him. And so off we went to la quinceañera to give La D-D some *mal de ojo* while she danced with Annette's future husband, though he didn't know it yet.

As we passed the mirror in the hallway, I remember thinking we looked so cool with our bangles clinking, and our matching black Gucci purses—okay, flea-market knockoffs. We were ready for Saturday night, decked out in our scandalous black skirts and our big white stretch belts with the shiny butterfly buckles. We sashayed our way to her mother's car, already idling in the gravel driveway, singing "Girls Just Wanna Have Fun." But when the car's headlights hit us right on, we had to stop. On cue, Annette and I looked at each other and performed an artistic interpretation of "Cherish" by Madonna. It was our own live music video, without the boy mermaids, an ocean, or good singing voices, that is. Her mom honked the horn and we laughed so hard, we shook like rattles.

We arrived just as the hall's doors were being held wide-open so a group of *señoras* carrying aluminum trays of enchiladas could make their way in. Annette and I had said good-bye to her mother and listened to all her warnings. "And, Lina, make sure Annette doesn't wander off with a boy. You stick with her, okay?" I nodded, Annette shook her head, and we walked in after *las señoras*. A

San Antonio summer night can slow you down like a long sauna session, but as soon as we escaped the smell of beans and entered the air-conditioning, we were instantly refreshed. Decorated with red, black, and silver balloons, the hall made us giddy with anticipation. Annette squeezed my hand as we stood before the huge archway of balloons framing the entrance to the dance floor.

They were already playing regular pop music from Duran Duran to Culture Club, so I knew we had missed the official waltz between la quinceañera and her father. I enjoyed witnessing that dance, because it reminded me that there are some real fathers out there. As I spaced out while staring at the cute DJ mixing his turntables, Annette tugged at my arm, signaling it was time to dash into the girls' restroom.

"How do I look? Did you see him? Should I wear the bow on the left side of my head?" Annette bombarded me with so many questions that she didn't even notice that La D-D was exiting one of the stalls behind her.

But I noticed, because she looked . . . well, she looked freaking gorgeous. The contrast of the white dress against her *morena* skin made La D-D look like a young Selena. Sexy. It was hard not to notice the silkiness of her cleavage, especially up against the white lace of her dress. The lace had patterned roses and little rhinestones all around it. Annette finally realized that we weren't alone, probably because she noticed my *cara de lela*.

"Hey, girl, you look so beautiful!" said Annette.

Annette never failed to amaze me. She can pour on the sweetness like she's a bottle of Ms. Butterworth. She even went up to La D-D and hugged her! So I followed suit and gave her *un abrazo*, fully aware that our chests had come in contact and feeling self-conscious about these thoughts.

"*Ay, qué bueno* that you two made it. I really don't remember who I invited. My mother did most of the inviting, and man, it's looking like a party for a bunch of *señoras*."

"Wow, your *vestido* is so beautiful," said Annette.

"Yeah. Ricky liked it too. Although I'm sure he's going to like it more when I put it over his head."

I started giggling. Out of nerves. Out of admiration because La D-D was so cavalier about sex. When most girls in our school were worried about being labeled as *putas*, this one flaunted it. But I also giggled as a way to distract La D-D from noticing Annette's expression. *Celos, mija.*

"Hey, don't we have gym class together?" she asked me.

"Yeah, girl, when you show up." I tried to be sly with her, have a moment of Chicana street repartee. She gave me a curious look with those almond-shaped eyes outlined in heavy black eyeliner. They made me wonder why we've never said more than a hello in the hallway to each other. But she'd never been in any of my honors classes, and I was too chicken to be as bad as La D-D.

Her crowd not only smoked the weed and did the dirty deed . . . they also didn't give a damn how many classes they got kicked out of. For all my *callejera* ways and smart-ass attitude, somewhere and somehow I got it in my head that my education mattered. Like I knew it was my only ticket out of San Antonio. I suspected La D-D didn't really care if she graduated or not. But someone must have cared, because a few years later she graduated right along with me.

"Well, have fun tonight," said the bad-girl quinceañera herself. My eyes caught one last quick glimpse of her cleavage as she sauntered out of the bathroom. And for a second, I envied Ricky.

Annette leaned in to me as we walked out of the bathroom and whispered, "She looks tacky with that whorish makeup."

"I think she looks pretty. Like a Mexican Cinderella."

"You would," she snapped back.

We found seats at an empty table and watched a smattering of little kids trying to dance. While I had inhaled about ten enchiladas, Annette only picked at her plate. She had her eye on Ricky sitting across the dance floor at la quinceañera's table of honor. I ate and scanned the room, hoping to see another familiar face, maybe someone else from school. I didn't. So I sat studying the seven *damas* in La D-D's court. They wore dark red dresses: strapless, extra tight on top, with a bottom half that billowed out past the knees. All of them wore their hair up, pinned with a red rose on the back, and a little Michael Jackson curl that snaked down on the left side of their foreheads. Depending on which girl you stared at, the look could be classified as "*elegante*" or "damn, girl." Two of the *damas* didn't have much of a figure to inspire even a lukewarm thought, but the older girls were voluptuous and made you marvel at the appeal of boobies. Some of the girls had those same almond-shaped eyes La D-D had; I figured they were cousins. Maybe sisters. I really knew so little about her.

But La D-D was right about one thing: The crowd was way over fifteen. Lots of *familias*. Some of the old-school *cholo* dads started undoing their ties, and exposing their blue-green tattoos from under their starched shirts. And as the Miller beer got guzzled, more and more sleeves got rolled up, and more and more brown and brawny forearms appeared atop the tables.

Annette and I were too chicken to drink too much for fear that our mothers would ban all quinceañeras, or worse, insist on chaperoning us. Eventually, once the "responsible" adults got too drunk, we ended up with beer bottles on our table. A sip or two never got us into trouble, just gave us that "we must be older than we look" edge at these parties. Made us feel like women.

I looked at the other women. Like my mom, these *mujeres* had that tired expression, the one they try to cover up with lots of heavy foundation. They smile, but the dark moons under their eyes betray them. Clown faces, I thought to myself. The *cansancio* was in their sagging bodies, too. With their men and babies pulling on them from so many directions, their flesh just loses shape. Like my mom used to say: *Todas güangas, pero buenas pa' las pachangas.* These type of women made me think maybe I shouldn't have sex until college, just because of the men and the babies that come along with it. Maybe.

Then all of a sudden I heard my favorite song by the Pet Shop Boys, "Always on My Mind." I had had enough of holding the table down with my skinny elbows thinking about my future. And if I ate another red enchilada I was going to pop out of my tight outfit.

I dragged Annette onto the dance floor and started lip-synching along with the music. And true to form, once Annette is on the dance floor she's my dancing twin, *mi otro yo.* Everyone always said we looked like sisters, except I was the much thinner and lighter version: *una güerita flaquita* and a late bloomer in so many ways. The music hypnotized our bodies and minds and we forgot about everyone and everything.

We are so into our moves, 'cause we are Chicana New Wavers. We're so cool we hold our hands behind our backs and spin like those beer bottles we twirl at house parties. ". . . You were always on my mind . . ."

That was living it up for us. Living it up. Even if it was only for one song. Even if it wasn't a song at your own quinceañera. Even if I had really wanted one, I knew my welfare *amá* could never afford it. Do people really know what kind of budget you need for one of these things?

Plus, the one time *mi amá* toyed with the idea—cause she was thinking of asking my rich uncle to pitch in and get lots of sponsors along with my *madrina*—I told her that I'd rather take the money we'd spend on one night and buy a '65 Ford Mustang. A red one. *Hasta convertible.* That way I could zoom my way out of San Anto' as soon as I graduated high school. "*Ay, mija, tu 'stas crazy sometimes,*" is all she said.

I think she was secretly relieved I wasn't going to begrudge her the fact that she couldn't throw me a *fiesta de quinceañera.* And maybe that's why she let me spend my summer weekends with Annette, crashing all of the quinceañeras that came our way.

Annette, of course, was having hers in a few months. I dreaded it. Friends had asked me to be part of their quinceañera courts before, but I always made up a clever lie as to why I couldn't participate: "Ah, you know what, Debra, I'm so sorry, but I'm gonna be visiting my cousin in Houston that weekend." "Oh, you know what I just found out, Patty, my grandmother needs me to take her to the hospital that day." "Mary, I would love to be in your quinceañera, but I . . . um, I have to . . . um, help my mom with something."

The quinceañera years were a blatant reminder of my family's lack of *dinero.* I knew that even being a *dama* in a quinceañera was just as burdensome on *mi amá.* Asking her to spend over a hundred bucks for a dress I was only going to wear once was asking her not to feed eight mouths for a week. *No creo yo.*

But I couldn't say no to Annette. I had a few months left to figure out how I was going to get money for that *pinche* emerald green dress she wanted all of us to wear. Even if it meant I had to get back together with my on-again-off-again high school boyfriend, Patricio. He had money, and good enough looks to be a respectable partner for the event. Like I said, I had enough time to strategize.

These were the thoughts that came to mind, as the Pet Shop Boys mixed into some Debbie Gibson or maybe it was a Belinda Carlisle tune . . . one of those blond chicks with that girl-next-door look. That is, unless you lived in our neighborhood. To us, next door meant *la comadre* Chucha and her ten chickens. Anyway, when I finally looked up at Annette's face, I saw that she was smiling wide, like when she got those sugar highs from gulping down cans of Big Red soda. She had forgotten about Ricky for an entire song. The DJ kept spinning one hit after another. After a string of George Michael, the Bangles, Lisa Lisa & Cult Jam, and the Miami Sound Machine, we were ready to switch dancing gears and cumbia to some Ramiro "Ram" Herrera, Texas Revolution, Selena, and our favorite, Emilio Navaira (who graduated from our high school, by the way).

More people clambered onto the dance floor, and then someone finally dimmed the hall lights. It was a signal that some serious partying was about to go down. For the few hours that followed everyone conveniently forgot that only thirty feet away from our hall of sin was our St. Lawrence church with its dark stained-glass windows of saints and Jesúcristo. But we knew our sins would be forgiven anyway.

And speaking of sin, she bumped into me on the dance floor while I was tejano dancing with some random *papasote* to La Mafia's "*Si Tu Supieras*."

"*Ay, mija*," she yelled over the music. "I'm so sorry. Did I spill some on you?" I checked my black skirt, and it was fine. But when I looked up to tell her this, my eyes landed on her cleavage instead. It was like a déjà vu. But, unlike La D-D, this older woman caught me looking.

"It's okay, *mija*, I'm sure you'll have some of your own in a few years." I must have blushed like a strawberry wine cooler, because she gave a reverberating laugh that made her chest heave. If she had been smoking, she woulda looked and sounded like a sputtering muffler. Then she spun away with her dance partner.

I looked up at the man who'd asked me to dance to see if he had heard the woman's comment about my underdeveloped *tetas*. With his thick caterpillar mustache and cowboy boots, he seemed oblivious to it all. Thankful for his *cerveza* buzz, I placed myself back up against his muscular chest and we drifted into the wonderful pulse created by the *gente* around us.

As he continued spinning me around, twirling me behind his back, I searched for Annette. I spotted her dancing with an older *tejano* as well. She was doing just fine. We have a signal, after all. "Pull the bow out of your hair if you need to be rescued," we said to each other before we walked in. But our bows were firmly in place. She didn't even seem to care that La D-D and Ricky were slow dancing in the center of the dance floor.

I searched for the woman who made me blush. It wasn't that big of a dance floor. Everyone was dancing so close to one another; I could smell the *mezcla* of Avon perfumes, beer breath, *agua florida*, cheap cologne, and perspiration around us. I didn't even know what color dress she was wearing, but even in those dim dancing lights, I had noticed the black beauty mark on her left breast. And that one-of-a-kind laugh.

I don't know why I wanted to see her again. It was like she reminded me of someone. She also wasn't like any of the other women there. She didn't look like a *señora mama* type; she was the *mamasota* kind. And then it hit me. *Why in the world am I thinking of this random woman when I am in the arms of a Mexican Marlboro Man?* I searched for Annette again, and when I didn't find her I excused myself from my MeChicano cowboy. "Don't run away," he said. "I'll be looking for you."

I made sure he didn't find me.

I headed toward the restroom; I needed to powder my nose, tighten my bow, and collect some good ol' Catholic guilt to curb my illicit thoughts.

There was a long line to the women's restroom, so I stood next to a *viejita* and watched the dance floor from afar. The kitchen door on the opposite side of the floor flung itself wide open to some obvious bustling going on in there. I heard a loud crash. Both the *viejita* and I jumped. She with her holy *"¡Diosito!"* and me with my more vulgar *"¡Chinga'o!"*

A couple of us walked over to the kitchen door and peered in to make sure nobody was hurt. An older man asked, *"¿Todo está bien?"* Someone had made a huge enchilada mess. They were everywhere, like red lava on the white tile floor. And then a woman with a bunch of paper towels came into view and yelled back, *"Sí, papá, no te preocupes."*

It was *the* woman. I stood there watching her as she bent over and piled globs of enchiladas onto silver pans, her breasts nearly spilling out of her dress. She caught me again. Before she could say a word, I blurted out, "Do you need some help?"

She looked me up and down and then smiled. "No, *mija*, I clean up my own messes. Besides, you don't want to get your cute outfit dirty."

Like a South Park Mall store mannequin, I stood and observed this woman grab handfuls of red enchiladas with her long, elegant fingers. Her breasts jiggled the faster she cleaned. That black beauty mark of hers moving back and forth like a metronome. I could feel the perspiration gliding down my neck like a wet tongue. I wanted to help her, but I felt all weird. I knew I wanted to get close to her for all of the wrong reasons.

Please understand, I used to go to catechism classes; I knew all of the lesbian nun jokes we used to crack about the sisters in the parish office; and I knew it was wrong, according to my religious upbringing, to deeply wish she would ditch those enchiladas and grab my hair between her fingers and . . . and . . . you know . . . plunge my innocent *carita* into her boobies! *¡Jesúcristo!*

Now that's a baptism for you.

And since this was the first time I'd had these wondrous thoughts about an older woman, if my wish had come true, I don't know what I would have done with her enormous breasts in my mouth . . . but I'm sure I could've figured it out. Call it instinct. Call it natural. Call it lesbian ancestral memory. This woman had this fierce sensuality, and a way about her that made me think her body provided all of the hands-on instruction anyone would ever need. Let me say it again, she wasn't like the other women in the room that night or any I had ever noticed before.

"So are you a friend of my *niña?*"

I didn't answer at first. I didn't realize she was talking to me.

"Are you okay, *mija?*"

"Uh . . . who me?"

She laughed again. The same sound that caught my ear on the dance floor, muffler style: guttural and sputtering. Then she wiped her hands clean with a wet towel. And then it occurred to me that she was related to La D-D, maybe an older sister, or a *tía*, because she didn't look that old.

So, I asked her if she was La D-D's older sister. And this pleased her, made her smile. Big-time.

"*De veras*, you think I could be her big sister?" She seemed to have a thought that made her go pensive, maybe even a little sad. And then she grabbed a half-empty beer bottle from the counter and walked toward me. I felt what Don Diego must have experienced when that powerful brown *diosa* appeared to him in all her glory. If this woman had said, "Get down on your knees," I would have. The closer she came, the quieter the room got. All I could hear was my own heart flipping, like a ball of *masa* in my *amá*'s tortilla-making hands. On her way out of the kitchen, she passed by me and caressed my cheek and said, "That's sweet, *mija* . . . that you think I'm that *jovencita*."

Just like that. Her hand singed my skin, and I was branded for life. I wanted more of this female kind of touch. I wanted that woman. I wanted to thank La D-D for turning fifteen and having the best quinceañera party ever. I needed to befriend La D-D.

I don't know how much time had gone by in that kitchen when Annette finally found me. "Lina-lu! Guess what just happened? What are you doing in the kitchen? Oh, my goodness, you're not gonna believe it!" I looked at my best friend, and for some reason, I couldn't even begin to explain what I had experienced. How I felt awake for the first time. I mean . . . would she have freaked out? I was freaking out.

"What happened?" I asked.

"Ricky and La D-D just had a huge fight! In front of everyone. Didn't you see him leave? Look, look over there. Her mother's trying to make her feel better."

And there was La D-D crying at her quinceañera table with *the* woman. My woman! I felt so bad for La D-D, and honestly, I felt a little jealous. *Celos, mija.* There was La D-D with her head on the woman's breasts. Everyone was staring, even those pretending to dance.

"That's her mom?" I asked.

"Yeah. Why?"

"She don't look like a mom."

"She looks kinda slutty."

"No, she doesn't! God, Annette, why are you always so judgmental of people."

And that was it. Annette gave me a look like I'd slapped her and she walked away. I watched her leave through the side door. I wanted to go after her and say, "Look, I'm sorry, but you're bitchin' about someone who just made my panties wet and I don't know what to do about it!"

But I let her go. She was undoubtedly going to sniff around for Ricky.

I glanced back at La D-D and her sexy mother. Her mother had pulled her daughter onto the dance floor. A cumbia was playing and I watched them go around the room in circles. It wasn't hard to spot the similarities between mother and daughter: the shape of the eyes, their jet-black wavy hair, the color of their skin, and yeah, that compelling cleavage.

I wondered about their lives. Somehow at that moment, la quinceañera looked more vulnerable and beautiful to me than ever. I felt a tinge of guilt for never getting to know La D-D all of those years. I had written her off like everyone else. But that so-called troublemaker and resident bad girl had a story like the rest of us. And her mama had stories for sure!

It occurred to me that I hadn't seen La D-D's father. She had to have had one to get her quinceañera through the church administration and all. Most of us kids, however, were part of the 1980s phenomenon: decade of the soaring divorce rates. But, trust me, I wasn't feeling self-pity because we didn't have a father around, especially the one we used to have, *tan borracho y macho*, he made the *vato* on the *lotería* card look harmless. In fact, my traditional *amá* getting a divorce, despite the cultural mores and the "*Ay, qué* shame"s from our *familia*'s hypocrites, made her very modern and revolutionary in my eyes.

It was evident that something revolutionary and *bien* rebellious made La D-D's mom tick too. I wanted to live that. Wanted to know it. Wanted to taste it. I was hungry and ready to seek out actual *mujeres* who walked a different *camino* with a "*¿Y qué?*" attitude. This was San Anto' after all, there wasn't much excitement in our conservative Catholic city. Remember how everyone was crucifying our mayor Henry Cisneros for his extramarital affair? Even as a teenager, I already knew, relationships were more complex than my Rubik's Cube.

The DJ played a slow song, "Hold On to the Night," the one by Richard Marx that makes you want to cry even if you and your love have never been separated by an ocean. That one. Well, La D-D's mom danced yet another song with her daughter. I've never really seen this at a quinceañera party, but it seemed so much more appropriate than the typical father and daughter waltz. I mean, shouldn't it be our mothers announcing our womanhood to the world?

La D-D nodded her head while her mother looked as if she were giving her some *consejos*. And by the next song, La D-D was all smiles and danced with a *señor*, probably one of the *padrinos* of *la fiesta*.

Then, like a good hostess, the mother started to make her rounds, visiting tables and cajoling guests to get up and dance. And that's when I saw it. I saw how the men at the table devoured her, and how their wives responded to their husbands' obvious attraction to this woman by rolling their eyes at one another and making rude *gestos* behind La D-D's mother's back.

Back then, my mom was not anywhere as gorgeous as this woman, but she was pretty when she was young. I've seen the pictures of my mother's voluptuous body and her mane of curly brown hair. In her forties, she looked like all of the other *señoras*, with the exception that she had managed to keep a mischievous and playful spirit alive in her. I think of her when I see the two-faced nature of *las señoras*. Quite often my mother warned me,

"*La envidia, mija*, that's what you got to watch out for in this *vida*."

Eventually, La D-D's mom made her way to my table. I was sharing it with a family of five that included a feisty three-year-old toddler. I felt embarrassed that I was sitting all alone, lost in my thoughts, shooing away the occasional drunk *tejano* who, like a *mosca*, insisted on buying me a drink. Never mind the fact that I was jailbait.

"You're not dancing anymore?" she whispered in my ear. Her warm breath on my ear sent a surge of heat down my spine. I wanted to faint. I watched her with my peripheral vision. She probably thought I didn't hear her over the music, so after she took a sip from her beer bottle, she leaned in and whispered to me again. The second time around I felt the silkiness of her cleavage lightly rub against my upper back. It took everything in me not to turn around and kiss this woman.

I don't know what would be more of an *escándalo*: two females kissing in full view, the obvious age difference, or the fact that I had the hots for la quinceañera's *mami*. Whatever it was, it was an "*¡Ay, Dios mío!*" for sure and too many "Our Fathers" to ever be forgiven by the priest.

I would have been willing to risk eternal damnation if I knew the feelings were mutual. Risk is now my trademark, especially among my more straightlaced friends. I'm now *la loca*, the one who ventures and explores, because I feel like it. Back then, however, I felt like I was going crazy having these desires and thoughts. Eventually, I would tell Annette everything that had transpired that night. I had to tell someone.

As for La D-D's mom, I guess she just walked away because I was too *gallina* to turn around. She depleted me of all words, made my tongue shrivel up like a dry *tamarindo*. I just remember sitting there like a stunned *mensa*, long after she had moved on to the next set of tables.

I was afraid that the slightest adjustment of my body would erase the tingling sensation she left in my ear. So I sat, taking in this first lesson of appreciating a woman's effect on me. I luxuriated like a lizard in the *sol*, pondering the moment's subtlety and danger . . . all of it so absolutely magical and naughty. I don't even remember hearing the music.

What I do know is that at some point—out of nowhere— a *viejita* yelled from the side door, *"¡Un pleito! ¡Jesúcristo, es un pleito! ¡Socorro!"* Half of the room was oblivious and too drunk to hear the old woman's pleas over the music, but those of us sitting close to the door rushed toward the parking lot.

Outside was pure spectacle: The WWE starring Ricky and some forty-year-old guy wrestling on the cement floor of the parking lot. Their high-class tuxedos were ripped and there was blood on Ricky's shirt. A church security guard appeared out of nowhere and attempted to break them up. Annette was off to the side crying and screaming, "Leave him alone!" And before I knew it a ring of bystanders gathered, including la quinceañera and her mom.

Two men from the party helped the security guard pull those Lucha Libre stars apart. Lots of cursing and bravado between the men continued while La D-D and her mom rushed over to the older man who had been fighting with Ricky. *El señor* looked pretty messed up, all gangly arms and legs, as the women held him up like a giant *títere*.

I walked over and comforted Annette, who was all *mocos* and drama. "Annette, what happened?" But she just shook her head. Finally, with good ol' *resentimiento* she blurted out, "What do you care?" I almost wanted to leave her there. We were not having a good night. We rarely argued as best friends, but when we did, we were like kid sisters who can't stand each other. But I stood next to her anyway. Even if she hated me.

Some of Ricky's younger friends arrived on the scene. But nobody started up any more trouble, because within minutes more security guards had appeared as well. And then La D-D walked up to Ricky and cussed him out. "You're such a fucking asshole, Ricky! It's my quinceañera and you do this to me?!" She gave him a good push. He lost his balance. Her rage was palpable, and her tiara wobbled on top of her head like a clumsy waiter's tray about to topple.

And then it happened. La D-D turned to look at Annette and there was this crazy look in her eyes. Annette became a bullet, and I was a human shield between La D-D and my best friend. "You want my man, bitch? Come and get him!" screamed La D-D.

Under normal circumstances, my little *cholita* ass would've just taken my heels off and started to defend Annette at all costs. But I just stood there, caught between these two girls, hoping Annette wouldn't further provoke La D-D.

"Fuck you! Ricky can kiss who he wants!" yelled Annette, who was pushing up against me as I tried to hold her back. Luckily, La D-D's mom intervened. She grabbed La D-D swiftly by the arm, spun her, and gave her daughter a quick and tiny slap on the face. It stunned La D-D; it silenced all of us.

Nobody slaps a quinceañera.

I don't even know if La D-D's mom realized what she had done. She just started drilling into La D-D in Spanish. Saying things like *I told you he was no good, but you're stubborn*, and *Look at what he did to your uncle*, and *I'll be damned if I ever sacrifice so much time and money for you ever again*, niña.

I didn't want to stare, but I couldn't pull away. La D-D's eyes were filling up with tears as her little hands clenched up into fists. I knew that homegirl hated her mom right then and there. We all knew that feeling. But even our bad girl knew not to mess with her mother.

At that moment, I lost all sexual attraction for La D-D's mom. It was drained out of me with the fear. All those warm sensations down there had turned into polar ice caps. La D-D's mom went all Freddy Krueger on us; her face became riddled with worry lines and her long fingernails became menacing blades. I wanted to scream, "Run, girl!" But La D-D stood her ground no matter how loud her mom got. In my head I was singing Lauper's "True Colors," hoping La D-D would keep it together. *(You with the sad eyes, don't be discouraged . . .)*

I guess I always knew that she and I could never be friends, not when she and Annette were rivals. Annette grabbed my hand. "Let's get out of here." I looked at Annette's runny nose, her bow all *chueco* and crooked on her head. And as we walked toward the outdoor pay phone to call her mom I needed to know one thing: "So is he at least a good kisser?"

She placed her head on my shoulder as we walked, and sighed, "I'm so in love with him, Lina-lu."

After we called her mom to come and pick us up, Annette and I sat inside the wooden gazebo in front of the church. Annette told me every little detail about the moment that led up to the kiss with Ricky, and how La D-D's uncle caught them and he called Ricky a *puto*, and how it all got crazy.

But I was only listening half-heartedly, until Annette started gossiping about La D-D and her mom.

"You know what Ricky says . . . that she's a lesbo. A fucking *marimacha*, can you believe that? That's so gross. He says that's why her parents are having problems."

"Who? La D-D?"

"No, stupid, her mother! Aren't you listening to me? I said that Ricky told me she has a *novia* on the side that looks like a man. I mean, that's so weird, why not just get a real man?"

I was grateful we were in the embrace of darkness, because while Annette went on and on about how wrong it is to be a gay, 'cause the Bible says this and that, my whole body smiled. A body can do this, you know?

It didn't matter if I never saw La D-D's mom again, she had given me more than she could ever imagine. She gave me hope that I would find others like me. Hell, I just hoped I could find more women like her. And then I decided to challenge my best friend on her homophobia and double standards.

"You know, Annette, the Bible says it's also wrong to have sex before you get married."

"Yeah, but that's different. Doesn't it gross you out?"

It was a moment of truth waiting to happen between two best friends. I thought to myself: Does she really love me? Do I trust her? Do I know what I'm saying? After all, I just have these feelings, it's not like I had ever kissed a girl . . . yet!

"Here comes my mom!" she yelled.

I was saved. We got up and walked slowly out of the dark gazebo. "Hey, isn't Sandra's quinceañera two weeks before school starts up?" she asked. I shrugged my shoulders, not really caring if I ever went to another quinceañera again.

Sure, you could say La D-D got to come out to the world as a burgeoning young woman that night, but I got to come out to myself. It was a baby step. I wouldn't officially come out until I was twenty-two. Before then I would try to push the experience of this quinceañera away, shelve it as an anomaly in my hypersexual hetero lifestyle.

But there's this beauty in living that's somehow like those *tejano* spins and the way they circle back into one another. The thrill of the return. For me the dance of life is not about how many partners you dance with, or how well you dance, or what style you dance . . . just that you dance long enough to see some of those partners return to you. Some changed, some not.

It is almost a decade before I see la quinceañera's mom again. At twenty-three, I returned one last time to the St. Lawrence dance hall for a wedding of two good high school friends. I still looked like a teenager next to all of my old friends, because whether it's San Anto's flour tortillas or the babies they're all making, they looked like *señoras*.

Annette went from Betty Boop to Betty Droop. She and I stopped being best friends long before we graduated high school. And my not participating in her quinceañera in the end probably didn't help our tenuous friendship. At the wedding, we were sweet to each other, in that superficial way you can be with someone you once intensely loved and cared for.

The queerer I got, the smaller and more provincial my hometown felt. I remember leaving my friends' wedding early. I was too liberal, too Californian, and too skinny for them. Trust me, they make you feel guilty for being thin. Eat! EAT!

Later that night, I found myself with some queer friends at the Petticoat Junction. Nothing like *jotería* two-stepping to *norteñas* to make my *tejana* self feel at home.

When I caught a glimpse of her, at first, I thought I was buzzing. In my early twenties, I was drinking alcohol like it was the holy water that would save me. I asked my friend José to confirm what I saw: A gorgeous older woman at the end of the bar. A beauty mark on the left breast. She was joking with the female bartender when I heard that undeniable muffler laughter rip through the calm of my body. *Boom!* Like a gunshot. I nearly jumped off my stool.

I'm not one to kiss and *chismiar* . . . but, yeah, la quinceañera's *mami*—she finally grabbed my hair like that mess of enchiladas and plunged my not-so-innocent *carita* deep into those boobies. I was born again, baby.

And I know that you must be feeling . . . *celos, mija.*

Barefoot

BY Eric
Taylor-Aragón

Many times people ask questions and they really don't give a damn about the answer. They're just filling the air with noise. This is what's called phatic speech. "How are you?" they might ask. Or they might say, "Nice day, huh?" or, to be more urban, "Whassup?" Phatic speech is a method of establishing presence, contact, and at the same time not really communicate anything meaningful except for the fact that you've been properly socialized. That you aren't going to pull a knife and sink it into someone's throat.

It was the summer of 1992, I had just turned twenty, and I was working as a carpenter in the Berkeley Hills of California. The year before, several thousand houses had been lost in a big fire, and there was lots of construction work. I had one more year at university and I was broke. I lived in a small room that I shared with a friend in a boarding house on Shattuck Avenue. Above us lived a bohemian painter type. Beneath us, in the basement, there lived a group of skinny, polite white kids who would later become rock stars and make millions.

The Berkeley Hills were alive with pounding hammers in those days. People had received huge insurance settlements and were building mansions, lots of them quite ugly. We were working on a massive Spanish hacienda–style house. Our crew was comprised of ten guys. Most were white Americans, then there was me, and about three or four Mexican laborers.

Antón was one of the aforementioned laborers—what stupid white journeymen carpenters called "*Manuel* labor." He'd dig the big trenches for foundations, he'd pick up the rubble, move lumber and bags of cement, sand walls, do cleanup. He'd do all the dirty work. He got paid ten bucks an hour. The contractor and the foreman were pretty decent folks, but Antón didn't have his working papers and, more important, he didn't speak English. I spoke Spanish, so I'd often have to tell him what to do.

It was Antón who indirectly caused one of the most traumatic nights of my life. He is to blame. But I can't be angry, because I think there are certain moments in life where trauma is needed, when disaster and mayhem become necessary. Antón was merely the vehicle.

Strangely enough, he was an individual absolutely without malice. A real gentle person. He was never sarcastic, never cynical, never ironic. He always asked questions with a freshness and resolve that made you think, "Man, this guy is really interested in my answer." He'd ask you how you were, or whether you had a good weekend, and then look at you with those hopeful quizzical eyes and actually be concerned about your answer. Phatic speech didn't exist for him; there was always something at stake.

Or maybe you'd mention a pretty girl you saw walking down the street, and he'd ask you what she looked like. Suddenly it became a sort of responsibility to explain her appearance to him in detail. He assembled the girl in his mind's eye as you described her. If you left something out, the woman might end up without eyes, or with no legs or bald. You had to describe everything, and only when you had done a very thorough job describing her would

Antón nod his head and say, "Yes, she's very beautiful. You're right."

I remember once we were on our way to the dump and drove by the oil refineries in Benicia. They were busy belching out their smoke in big silvery puffs, and there were a few scattered cotton-ball clouds in the sky, and I told him that the refineries were actually cloud factories. That because of global warming, et cetera, scientists had devised a way to make clouds artificially, and he was amazed, "But that's amazing!" he said. "Incredible! Crazy gringos!"

When I finally told him I was lying he looked at me with those big dark eyes and I think there was something deeply pained in his gaze, but after a few seconds he started laughing, a big belly laugh—and then he rolled down the window, as if there was suddenly a very bad smell in the car. I felt bad then and apologized. *I must try to be more truthful*, I thought, *I must try to be a better person*.

This is important to me, this whole thing about being a better person. Let me try to explain. You see, fundamentally, I'm very shy. And I often feel a tremendous divide between myself and my fellow humans, and to close this gap I make terrible jokes or tell wild lies as a way of catapulting myself into a different reality, establishing a new dynamic. Or maybe I'm just an asshole, I don't know. In any case, when I told Anton this lie about the cloud factories, half of me was thinking, *Wouldn't it be interesting if this were true!* and my other half was thinking, *What a sweet, gullible guy this Antón is. What a tremendous capacity for wonder!* My capacity for wonder had steadily diminished in the last several years. Sometimes I thought it was gone altogether. I don't know if this was simply part of growing older, or born of a difficult family history, or maybe just too much post-modern philosophy—but it was true.

When Antón found out about my lie, he was pretty disappointed—but not only because I'd lied to him; he was also disappointed because the cloud factories were fictional. They didn't exist. So basically, an alternate reality was created, and then, a few minutes later, destroyed. A voyage! And if I was filled with wonder at some point, it was a wonder at Antón's wonder.

One day the foreman sent me and Antón to go pick up a custom-made door from a warehouse in West Berkeley. It was a hot day, so this was a welcome break. On the way, we stopped and got iced coffee and scones from Peet's. We sat on a bench beneath an oak tree and enjoyed our little respite.

After a while Antón turned to me and declared, "Enrique, I have a question."

"Go ahead," I said, taking a sip of coffee.

"When you look into a mirror your right hand turns into your left hand and your left hand turns into your right hand, no?"

"Yes . . ."

"But if you turned the mirror on its side, shouldn't your reflection be upside down?"

I had no idea. *"Buena pregunta,"* I said.

Ten minutes later, when we were back in the truck and I was pulling out of our parking spot, he turned to me again. "Enrique," he said, "I must ask you something."

"Well go ahead and ask."

"It is very important to me, Enrique."

"Okay."

"What are you doing Saturday?"

"I don't have any plans yet."

"I would be honored and grateful and very happy if you would come to my cousin Andrea's quinceañera celebration. She is a very special girl. I want you there. You will be my guest. There will be very good tequila. Will you do me this honor?"

And then he looked at me with those big chocolate-brown eyes. What could I do? I said yes. *At least there will be good food*, I thought. I'd been living on lentils for weeks.

"Thank you," he said. "This means so much to me. I will call you tonight with the directions to the church."

As it turns out, I missed that part of the celebration. The next day I woke up with a sense of foreboding. I'd slept badly, thrashed about all night, tormented by strange dreams. I felt disoriented. The ceremony was taking place in a Catholic church down on Martin Luther King Way —but I guess I must have driven right by it. I drove around, got thoroughly lost—and when I realized I was almost an hour late, decided to cut my losses, relax for a while and then go to the address that Antón had given me for the party.

I had a beer in Teacakes Dancing and Sports Lounge, bought myself a new shirt over at Baxter's Sophisticated Fashion, and took a stroll around the neighborhood. Before long, I found myself wandering by The Apostolic Church of Deliverance, a long, low-slung building with some busted-up bushes out front. A big black man in a suit stood by the door. I nodded to him, he nodded back, and I walked in. I sat down in a pew near the back. The Reverend Alfred T. Lexington was giving a sermon about calamity. He had a big deep booming voice and he almost sang as he spoke, *"Job arose, rent his robe, and shaved his head and fell on his face . . . Shortly after this he contracted a terrible skin disease. But he never lost faith, oh no he didn't!"* After this most excellent sermon (on a Saturday!), I felt tremendously hungry and went over to a local soup kitchen on 31st Street but it was closed and there was a sign on the door that said, DUE TO CIRCUMSTANCES BEYOND OUR CONTROL THIS

FOOD PROGRAM WILL BE SUSPENDED UNTIL FURTHER NOTICE. So I
went hungry. I figured it was for the best, since I was going to be
eating piles of Mexican food shortly. I looked at my watch. It was
time to go to the party.

I rolled down Adeline Street and saw a big street sign that
said:

WARNING TO MILITARY MEN:
DO NOT SHOOT DOWN UFO'S
YOUR GREAT-GREAT-GRANDCHILDREN
MIGHT BE IN THEM

I don't know what this means. I don't even know why I'm
mentioning it here. I kept driving. (I keep saying "driving," but
what I mean is *riding*. I was on a Schwinn ten-speed which I'd
bought at the flea market the year before. I want to be totally honest here.)

Anyway, some good homestyle cooking, a quick exit, that was
my plan. I still wasn't quite sure why Antón had invited me. I
mean, from what I understood, quinceañera parties were usually
pretty tight-knit family affairs. I'd borrowed a tie from a friend
and had ironed my favorite pair of pants. Originally I'd been
wearing a very boring red polo shirt, but after my visit to Baxter's
Sophisticated Fashion I was pimping a brand-new (80 percent
polyester, 20 percent rayon) shirt, sky blue with a white seagull
motif. I looked good. I skipped the tie. I was feeling sky blue.
Monochrome. It was late summer, a cool breezy Northern California day. My skin was pretty dark from working outside so much
and I'd recently bought some very fly white suede shoes with gold
buckles that offset my tan nicely. I also had some Ray-Bans on,
Erik Estrada style. The overall effect, I think, was one of re-
strained elegance.

Finally I found the house. The neighborhood was not the best I'd ever seen. This was not North Berkeley. There were no wine bars for miles. No exotic bakeries or aromatherapy shops. No pseudo-hippy youth from the suburbs panhandling. In fact, it was a bad neighborhood, not the sort of place I wanted to be after dark. A police helicopter buzzed overhead. I locked my bike out front.

The house was an old white Victorian. It was protected by a tall wood fence with bright glittering pinwheels lining the top, spinning like crazy in the light breeze. Inside I could hear people talking and laughing, James Brown singing "Try Me." I walked through the front gate and out of nowhere a very big dog lunged at my person. I leaped back, slamming the gate shut, and nearly fell on the sidewalk. This was an ominous beginning. If I'd been smart, I would have turned around right then and started running.

I heard someone laughing. I turned around and there was Andrea. I mean it had to be Andrea, right? She was wearing a white chiffon dress and a tiara. She was pretty, medium height, with sparkly mischievous eyes, hair tied up in a bun, lips painted a bloodred, and a slightly big nose that gave her a more mature, pensive look. The dog was barking like crazy and I didn't really know what to say so I said, "Hi, I'm Enrique, Antón's friend from work."

And she said, "Oh, so you're the one!" and started to laugh again. Then she turned toward the barking and yelled, "Shut up, Bruno!"

Behind her was a slender, haggard-looking fellow with droopy eyes and a little fu-manchu goatee. He was wheezing a bit. "That's Uncle Diego. He has emphysema. He was coming out here to get a hit off his oxygen tank but I just caught him smoking a cigarette." She turned to her uncle. "Now why were you doing that? You wanna die before I turn sixteen?"

"Nice to meet you," I said. "I'm very sorry I missed the ceremony. I couldn't find the church."

"Oh, don't worry about it. It was silly." The dog kept barking. "Why are you wearing all blue? That some sort of fashion statement?"

I blushed, and then immediately thought, *Why am I blushing? She's fifteen years old! What does she know about fashion?* She looked at me keenly, then gestured toward the barking, "That's Bruno. He's probably the only one in my family who understands me, except for my dad, who doesn't talk much. Anyway, let's go in. I have to introduce you to Mafalda."

"Mafalda?"

"Yeah, Mafalda, my aunt. That's why you're here. Didn't you know?"

I followed her in. Bruno stopped barking and jumped on Andrea. He was a mean-looking beast, a Rottweiler with a big scarred-up head, a tremendous Paleolithic jaw, and a fierce set of teeth. Andrea stepped to the side, gave his big mean head an energetic shove, and cooed, "Hush, baby, *niño, ¡cállate ya!*" and the dog shut up, its tail wagging like crazy. My heart was still pounding. I tiptoed by him, following Andrea and Uncle Diego to the house. Bruno bared his teeth, his eyes turned to slits and he crouched, as if about to leap. He gave a low rumbling growl. I froze. Andrea turned and yelled at him again. "Bruno! Be nice!"

We walked up the wood stairs, onto the porch, and into the house, Antón opened the door and he was dressed all in white and looked a little tipsy. When he saw me he said, "Ah, great to see you, *hermano*, let me go tell Mafalda!" And he disappeared.

There were lots of people milling about. I suddenly felt bashful and out of place. Perhaps I'd overdressed. I sidled into a corner of the foyer and looked around. There were pink crepe streamers swooping across the ceiling, and over the double door to the living room there were some letters that read:

¡FELIZ QUINCEAÑERA ANDREA!

The house was filled with balloons of all different colors. Every now and then one of them would pop and it sounded like a gunshot. There were dozens of screaming children careening around whacking at the balloons. The hall was packed as was the kitchen.

Andrea appeared with a woman I took to be Mafalda. She was very plump. Very, very plump. *Gorda* even. I am being too polite. My God, the woman was huge. Anton was beaming. "Mafalda, *mi amigo* Eric."

"Eric, *mi tía* Mafalda!"

"Encantado," I said.

"Mafalda is visiting from D.R.," Anton said. "She's single."

He must be out of his mind, I thought to myself. *Is he trying to set me up?*

Mafalda reached out toward me, and I thought maybe she wanted me to kiss her hand but then I realized she was holding out a shot of tequila. I grabbed it and downed it quickly. *"Gracias,"* I said, wiping my mouth with my sleeve. Andrea, who had been witnessing all this, started laughing. She laughed like Antón, with a hand on her belly. I remember thinking she looked younger than fifteen. Suddenly she stood up and said, "I'm going upstairs to change out of my dress." And with that, she was gone, and I was alone with Mafalda.

I could hear Santana playing in the living room. *"Oye como va."* Mafalda looked at me. She had light brown eyes, full lips, great round cheeks, large gold hoop earrings and her hair tied up in a messy bun. She had a double chin. Maybe triple. She was wearing a white dress embroidered with flowers. Big, bright, bodacious flowers. Tropical. Like a botanical garden. She licked her lips and looked at me as if I were some sort of succulent dumpling.

"Antón has told me a great deal about you."

"He has?"

"Yes. That you're Peruvian. That you're a writer. I really like your shoes, by the way. Suede, right? Very sexy. I love to read, you know. There is nothing I like better."

"Really? What do you like to read?" I asked.

"Books," she said.

And then Mafalda started laughing. It was a strange thing, this laughter of hers. It started out with several rapid, powerful exhalations, like a steam train picking up speed, then she went silent, her mouth open but with no sound coming out, her prodigious flesh trembling, her eyes clenched shut in glee, her chubby hands raised in the air and all the fingers fluttering and waving about like sea anemones, and then, as if someone had turned up the volume, a noise started to come out of her mouth. At first it was faint, but gradually it grew raucous, wild, uncontrollable. A tremendous earthy chortling. I could almost swear that the floor shook a little. It was like standing next to a volcano. She stopped laughing, looked at me panting, and spoke: "So, college boy, you like older women?"

I looked at her for a moment, stunned, and then Antón appeared, holding three more shots of tequila. "Come on, Eric, what are you still doing here? I have to introduce you to everyone. Mafalda can wait."

I downed the shot of tequila.

So did Mafalda.

Antón took me by the arm and led me to the living room.

"What's up with your aunt?" I asked.

"Oh, Mafalda? She's great, isn't she? *Una loca. Una belleza.* Don't worry, you'll have plenty of time to get to know her. Relax. Go with the flow. You'll see."

Antón proceeded to introduce me to Andrea's father and mother. I was still a little puzzled by his comments, but got so caught up in introductions that I quickly forgot about them. Casimero Dominguez was a slender man with eyes so dark and glinting the pupils seemed like black marbles with a spark caught inside. He had a moustache, a deeply lined face, thick Indian hair graying around the temples, a broad back, and a humble, dignified air about him. Then I met his wife, Angela, who was quite stout and fierce-looking, and wearing a bright embroidered Spanish shawl that fell past her waist. Wouldn't want to get on her bad side.

"Isn't our Andrea beautiful?" she asked me.

"Yes," I said, smiling, "she looks great."

I also met Andrea's brother, Ramón, a tall, serious-looking cat with a bit of an attitude. Antón had told me that back in Fruitvale, Ramón had been getting involved with the local gang and that's why they'd moved. And then there were several of Andrea's white friends, who were mingling about rather aimlessly, and then some more of her Latin friends from school, and they all seemed very young. The music was playing mostly old-school stuff, with a bit of Prince mixed in here and there. All the adults were drinking from the stash of beer and tequila in the kitchen. There was some pink punch floating around too.

A lot of people were in the garden, Antón told me, and so I waded through the balloons and went toward the back of the house. Andrea was there with a video camera filming everyone. She'd changed into jeans and a T-shirt and was going from group to group. She got to me and said, ". . . and this is Enrique, who works construction with Antón and who we invited so he'd get it on with Tía Mafalda, who likes *vatos jovenes*." And then she started laughing again. "*Ay*," she said, "what a silly party. I think I'm gonna leave soon. I didn't really want a party, you know. Mom made me do it. Do you like my video camera? Papa gave it to me; he knows I like movies and last year I took a little course, you know. It was the bomb. Anyway, this was his present, you know?

But he gave it to me yesterday so I could film the party. Pretty cool, huh? But I'm getting bored, all the grownups are getting drunk, all the girls are getting stupid. I'm gonna go."

"But you can't leave your own party," I protested.

"Oh yeah? Just watch."

"Hold on."

She looked at me seriously, "You know, this whole quinceañera thing is crazy, I mean, this is where I'm supposed to turn from a girl into a woman . . . and now supposedly, I can start dating boys. But boys are stupid, trying to be tough, buying the right sneakers. They're dumb. They're ghetto dumb. I mean *really*. I'm just not down. I wanna get out of here. I've never even had a crush on a boy. Can you believe it? I mean, this whole love thing and all these songs about how so-and-so can't live without so-and-so. It's all hype, man. All of it."

"Come on now . . ."

She spun around and disappeared into the throng.

People started to dance. Mafalda came to me, seized me about the waist, and before I knew it we were swaying along to the *ranchero*. I wouldn't call it dancing, it was more like being tied to some huge buoy just before a storm is about to hit, when the swells start getting bigger and bigger and the sky darkens and a wind whips up, and there's the sound of distant thunder getting closer and you don't know what to do, you're helpless, miles from land . . .

I inhaled discreetly. She smelled good. Like ripe apricots. She looked up at me with a charged gaze and declared, "I love America," and I said, "Really? And she said "Really," and by this time I was a little drunk and her body was big and warm. I'd never felt anything like it. Then she said, "I want to stay here, I don't want to go back to Mexico."

"How did you get here?" I asked.

"On a plane, *pendejo*."

"Oh."

"I love America and I don't want to leave. You're single, Antón tells me."

Antón appeared with some more tequila. I took one so I wouldn't have to answer her question. "Excuse me?" I asked, after I'd downed my tequila.

"You're single."

"Yes. So?"

"So marry me, baby."

I sputtered, lost my balance, staggered backward. She made a grab at me. "Just a minute," I said, smiling politely but firmly. She was about to say something else, but I slipped away. *I must be cautious*, I thought, *I don't want to offend anyone. I must be diplomatic, firm, keep things under control.* I walked back into the garden where some small children were banging away on a piñata with baseball bats. Didn't seem very safe. Parents looked on happily. Andrea was there, dancing around with her video camera. I liked her. Lots of energy.

I found a spot next to a cement statue of a bunny. I leaned on the bunny. I took my Ray-Bans from my pocket and put them on. There are times when I feel like a movie star, particularly after four tequilas. A light breeze was coming up off the bay. My new sky-blue shirt with the seagull motif was shining and fluttering in the wind. I started a conversation with a nice lady from Fruitvale who told me she was very into astral traveling. She also told me her cousin had recently entered California via the desert route at the base of the Chocolate Mountains and that everything that could go wrong did go wrong. He got bit by a snake, fell off a cliff, was mugged by *bandidos*, and got rounded up by the border patrol. But somehow he'd made it through in the end and was now at her house recovering from his journey. *"Tiene un ángel de la guarda,"* she said, nodding.

That's some guardian angel, I remember thinking.

Memories get vaguer at this point. I talked with Casimero, Andrea's father. "Because you see," he said, "Andrea's different. She's different, that one, and that's good. It takes all kinds, right? She'll go to college, be a doctor, or a moviemaker, or whatever. But she'll go. I'm proud of her. Real proud of her. She's a tough cookie. Got spirit. Stubborn as hell."

I copped a cigarette off Uncle Diego. "Enjoy!" he said, wheezing. I never smoke. Anton came up to me. "What's up, *hermano?* You having fun?" He looked at me searchingly. "Are you? Would you like me to get Mafalda?"

"No! I mean, yes, I'm having a great time. Oh, yes. So many new people to meet!" Anton winked. I wandered around the party. People were in the living room dancing to Joe Arroyo. There was a big table of food laid out. Tamales, chicken and beef enchiladas, wild rice, chips and guacamole, and corn on the cob. A good spread. Strangely enough, I wasn't really hungry. I had a glass of water instead.

Antón was dancing with a friend of Andrea's, really tearing it up. I felt strangely out of it. Like I wasn't part of the party, just witnessing it. In fact, I wasn't really sure why I was there. I mean, I was there because I was invited, but what did all this have to do with me? I went to the foyer and was about to open the door to the front porch when I heard a growl. Bruno was lying in front of the door. I turned around. There was Mafalda. I tried to play it cool. She winked. I saw Andrea come down the stairs. She winked at me too and went toward the garden. *What was going on?*

I went into the hallway, started talking to someone else, I don't know who. Maybe it was with her little brother. Yeah, that's it, Ramón. His pants were sagging and he was wearing a red bandana. Ramón was a good kid trying to survive in a rough neighborhood, a rough high school, and at that age where all the young bucks try to be hard. Suddenly he turned to me and asked, "*Oye, hermano*, you ain't a *sureño*, are you?"

"What?"

"You ain't a *sureño*, are you?"

"No, man, I'm from Berkeley."

"Just making sure."

"Making sure of what?"

"That you ain't one of them."

Now, I'd always heard about this great divide between the *norteños* and the *sureños*, but I'd never taken it seriously.

"And what would have happened if I was a *sureño*?"

"I dunno, man, but it wouldn't have been good." He smiled, flashed me a sign.

"Oh man, you gotta chill out," I said. "That's for fools, man . . ."

"I don't like *scrapas*, *vato*, what can I say? It's in my blood."

"Look, man, I'm half Peruvian, this *norteño-sureño* fight ain't my thing."

"All right man, jes checking. You talk like a white person, *vato*, anyone ever tell you that? No slide in your glide."

I was about to respond when I heard someone whisper in my ear, "Baby, you gotta try my tamales, they the best in the world." I look to my right. It was the volcano. Then I heard another voice. "Say it to the camera, auntie!" Andrea appeared with her camera, "Say it to the camera!" She was having a great time. As much as she pretended her quinceañera was stupid and threatened to leave it, she was really enjoying herself.

Mafalda looked at the camera, licked her lips, and said, "You

gotta try my tamales, they're the best in the whole wide world."
And then she licked her lips again and gave her hips a little shake.
The room trembled.

Trying to play along, I picked up a tamale. I unwrapped the
corn husk covering. Johnny Chingas was playing, "*Se me paró.*" I
took a bite. Suddenly an intense feeling of pleasure swept over
me, my mouth became flooded with cheering multitudes, my taste
buds bloomed into beautiful and delicate orchids, a baobab tree
sprouted from the crown of my head, a curious tingling spread
through my body, and I could almost swear that I levitated several
inches off the ground.

"It's very good," I said, "very, very good. Superb. Outstanding."

"I know," she said. "I told you, didn't I? You should try my
taco sometime, college boy . . ." Then she leaned in to a woman
next to her and whispered, just loud enough so that I could hear,
"My fiancé. *Mi prometido.*"

I turned to Andrea, who was still videotaping everything, and
said, "*Maravillosa.* Outstanding."

"You're crazy, Enrique. *Loco.*" And then she spun around and
started to dance with Anton.

The tight bun I had first seen her with had come undone. Her
thick black hair cascaded down her back and whipped this way and
that as she danced. She was at that age when the world is a vast ex-
panse, a mad wild infinite place with endless possibilities. Life
seems a huge and benevolent thing, a magnificent conspiracy designed to bring her laughter, joy, success. I'm sure if I'd met her
when I was fifteen I would have fallen madly in love with her and
she would have ignored me. This is a cruel truth and I was so af-
fected by it that tears came to my eyes. But only for a moment.

* * *

There was a commotion coming from the living room. They were bringing out the cake. It was a great big white thing, with a big rose made of frosting in the middle. People gathered around. Someone turned down the music. Andrea went to the front of the room and said, "Thank you so much to everyone, *gracias a todos por vuestra presencia . . .*" She was cool and collected, gave her speech first in Spanish, then in English, didn't use any notes, her gaze moved around the room from face to face. She smiled. This girl would go far.

Everyone clapped, sang *"Cumpleaños Feliz"* to her. She blew out the candles. More cheers. *"¡Belleza! ¡Guapa! ¡Divina!"* Her mother was weeping. Her father's eyes were shiny. She went to them, hugged them both. They put on a song, a slow ballad, and she danced with her father. She was a lucky girl. I hadn't seen my father in years. I wandered into the kitchen, had a drink of water, and tried the guacamole.

I asked someone for a cigarette and proceeded to smoke it. This was my second one of the night. I don't usually smoke, but there are certain moments in life when it becomes necessary. This was one of those moments. I saw Angela, Andrea's mom, and she asked me if was having a good time, and I said yes and smiled. I leaned against the wall, holding the cigarette between my third and fourth finger and smoked. I looked around. No one was talking to me. I suddenly felt like talking, and no one was talking to me. Things start to get fuzzy here. Like someone pressed fast forward. I remember fragments of conversations, some faces. I think one of them was named Ezekiel (You can call me "Easy"), this plump balding guy from Chihuahua, who told me he was an astronaut. His feet made little steps in time to the music as we talked. He was smooth, very chill. I could easily have imagined him on the moon, hanging out, leaning on his Chevy Impala.

"Oh, yes," I said, "very interesting indeed. I think Michael

Jordan would look good with an afro too. No, but really, this Chinese epidemic in the countryside is truly frightening. I mean, people's heads are literally exploding! Imagine you're talking to someone and their head suddenly explodes!"

Easy looked at me, perplexed.

"So how long have you been an astronaut?" I asked him.

"Who said I was an astronaut?"

I was making Easy uneasy. He drifted toward the garden. I found myself trying to make my way in the crowded hallway again. I was jammed against the wall, wedged into a corner by the exorbitant rear end of Angela, who was talking to a couple I hadn't been introduced to. Then I met a woman named Carmen and we talked about the recent invention of square watermelons. I was starting to have a good time. She wrinkled her nose. I smelled smoke. Yes, there was certainly smoke. Oh, yes, there was. It was getting thicker. I'd forgotten that I had a cigarette in my hand. I looked down. Catastrophe. My cigarette had burned a hole in Angela's shawl, which was now smoldering in a very alarming manner.

I've always had very quick reflexes, and without thinking, I immediately started to hit at the fire. In fact, I started spanking the prodigious rump of Angela with great energetic strokes. Formidable Angela. This, of course, caused a rather strong reaction; she yelped and spun around, but she spun so quickly that the fire flared up again, so I leaped behind her and smacked at the fire again and then again—and she yelped again. My God, what a mess! I was spanking the bottom of the quinceañera's mother! Horrified, Guadalupe threw her punch on me, and then Angela swung at me. And man oh man, she hit me hard, at which point someone yelled, "FIRE!" and Paco tossed a beer on her and Angela was soaked, dripping beer, and you could see her big pink panties through her dress. And as I staggered from the slap, I looked up just in time to see the blur of a fist traveling through the air and *BAM!* Right in the jaw. That's the last thing I remember. Apparently I just keeled over.

I woke up in Mafalda's arms, my head leaning against her tremendous bosom. I was surrounded by people, all of them apologizing for the big misunderstanding. Mafalda was holding an ice pack against my jaw; Joey, the guy who hit me, had brought me a shot of tequila. Very considerate fellow. *"Lo siento, compadre, es que no hablo inglés, no sabía que hubo un incendio, ¿sabes? Pensaba que estabas metiendo mano ..."* He shook his head ruefully. He handed the tequila to Mafalda, and before I could protest, she poured it down my throat.

I passed out again. I woke up on the couch and the party was in full swing. I'd been totally forgotten. My jaw hurt bad. My head felt as if someone were repeatedly hitting it with a ball peen hammer. I struggled to think coherently. After several minutes of painful cogitation, I decided that maybe it was time to make a discreet exit. So I struggled to my feet, went to the front door, opened it very gingerly, and stepped out onto the porch. It was dusk. Bruno pounced. I fell down, and scrambled toward the house. Bruno was in full attack mode, growling and baring his fangs. I tried to crawl away. He started tearing at my pant leg. I made it into the house. I slammed the door. But somehow in the excitement, I forgot my ankle was in the door.

What an astonishing pain. I screamed. My ankle felt broken in many places, as if it had just been slammed in a door. Bruno ripped at my pant leg, attacked my foot, tearing apart my brand-new white suede shoe, his head whipping this way and that. Perhaps he thought my shoe was some sort of wild animal, like a bunny or something. I screamed. The beast was going to eat my foot. I screamed again. They were singing *"Las Mañanitas."* Oh, I'll sing too, I thought, I have a good voice. Yes, a remarkable voice, many people have never complimented me on the beauty of my voice, but this is only because they are envious. Me, I'm not envious. Bruno continued to devastate my shoe. No one heard my screams. I couldn't get my foot inside the house or open the door wider because I was afraid Bruno would attack my face. I was deathly afraid of Bruno. Of his teeth.

I almost wept. *Perhaps I will have to sacrifice my foot*, I thought. Perhaps it is time. I'd just give my foot to Bruno. I read in the paper the other day that scientists had invented a chocolate that doesn't melt and I thought, *That's already been invented. It's called brown plastic.* I also read that more and more polar bears were being born hermaphrodites. Then, with a vicious snarl, Bruno tore off my shoe. I was free; he was there, masticating and slobbering on one of my shoes. I pulled my foot inside, slammed shut the door, fell on the floor, held my ankle. *I can't take any more. I'm finished. Dios mío.* Can't take no more punishment, I thought, weeping in my mind. My white suede shoe had been digested. My sky-blue pant leg was ripped to shreds. My sky-blue shirt with the flying white seagulls was stained with punch. My jaw was very swollen. There was a sharp throbbing pain in my head.

Mafalda came toward me with some cake.

"*Mi amor,*" she said, "*te he traido un poco de tarta.*"

Oh my God, come to me, Mafalda, come to me.

Then I saw Andrea come down the stairs. She stepped over to me, opened the front door. Just before she left she turned around and put her index finger to her lips. "*Shhh.*" I was confused. I thought to myself, why was she leaving her own party?

Everyone was in the kitchen, dancing, talking, laughing. I was invisible to everyone except Mafalda. She sat on the couch and patted her lap. "Come here, college boy," she said. And I have to admit that this is what I wanted most. To be held. Have you ever felt this way? Just wanting to be held?

My ankle felt busted. *It might be broken*, I thought. I was almost weeping. I picked myself up off the ground and hobbled to the couch, oh yes, to put my head on her warm bosom, that is what I wanted most, to be enfolded in Mafalda. I was tired, and I hurt everywhere. Mafalda fed me cake. She is a goddess. I didn't even have to move. I just opened my mouth. It hurt to chew but I chewed anyway. The people milled around, danced, flirted, laughed. The party went on. I ate my cake.

I must have passed out again. Several hours later I was roused from my catatonic state by the sound of someone's key in the front door. I was groggy. I peered over the top of the couch. It was Andrea. She tiptoed into the living room, took the video camera out of her bag, looked at me and mouthed, "Watch." She hooked the camera up to the computer. Mafalda was snoring up a storm. I looked at my watch. It was 11 p.m. Eydie Gorme was singing *"Piel Canela."* Andrea pressed REWIND. Blue screen. She pressed PLAY. I was delirious.

I saw myself at the beginning of the party, before I got so beat up. I looked alright. Healthy. I saw all sorts of people, her mom and dad, happy and smiling and saying *"Te quiero, Te queremos,* We love you, Andrea," and other folks, saying, *"Linda, guapa, preciosa."* I saw Bruno, looking into the camera happily, slobbering—his fangs were sharp, glistening, and yellow. I saw Uncle Diego taking hits off his oxygen tank, Antón dancing salsa with Mafalda. They danced great together. Then I saw Mafalda saying, "Baby, you gotta try my tamales, they're the best in the world." And doing her little hip shake. Andrea zoomed in on her breasts. Astounding.

Then I saw myself again, and then other images. The piñata,

children swinging baseball bats, the cake. I saw myself trying to put out the fire, getting slapped, punched and doused with pink punch. Oh that was a great scene. Andrea sat cross-legged on the chair in front of the computer, laughing her head off. She rewound, so she could see the punch again. Oh, that Joey! He got you good! *Bam!* (Later Anton told me he'd been a semi-pro boxer.) Then I saw myself falling over unconscious *"Dios mio,"* the women cried out, while the men carried me to the couch. More images of the party. Dancing, her brother flashing gang signs. Then I see myself slamming my foot in the door and screaming and getting attacked by Bruno. (Where had she been? Why hadn't she helped me?) I heard her laughing on the sound track as Bruno tore off my foot. She rewound the tape so she could see this section again. Then Mafalda woke up and watched the video too. They both laughed uncontrollably. It was great fun. We were having a blast. Andrea fell off her chair, she was laughing so hard.

The camera led us out the front door. Andrea left the party, swiveled one last time so that she got a final shot of the house. Then the bus came and she got on. I saw Oakland go by. This part of Oakland is a little sad: abandoned houses, projects, empty lots and pawn shops. She got off the bus, camera rolling, went to some café. It looked like Lakeshore, a nicer area about twenty minutes away. The café was mostly empty. It looked like her camera was in her bag or something, because sometimes the lens got partially covered and she had to adjust the opening. There was a thin man talking to a woman with green eyes and a wild head of curly hair. It was an intense conversation. They looked into each other's eyes. Some sort of fight maybe. Then a long gaze. His hand reached across the table. A pause. Her hand in his hand. Nice.

Then she zoomed in on a youngish-looking kid, probably sixteen years old or so. He was reading a book at a table on the opposite end of the café. He had thick eyebrows and a concentrated look about him. Every now and then he took down some notes. He glanced at the camera. Did he see it? He returned to his book. He took some more notes, no wait, he's drawing something. But what? After a while, he looked at the camera and then he looked away. He was in profile now. Nice profile, high cheekbones, strong face, disheveled hair. He was wearing a T-shirt and a secondhand leather jacket. He held up his drawing. It was of a big eye. The camera wobbled. The boy stood up and walked toward the camera until he was totally blocking it.

"Hello," he said, "my name is Marco. You're in my history class, right?" The camera turned off.

The camera turned back on. It was Marco again. A close-up. Shoulders and face. He was standing against a wall. It looked like they were still in the café. Andrea's voice: "Say it again."

"Hello, my name is Marco, and I think you're very strange and beautiful."

Marco was blushing wildly.

"Say it again!"

"Hello, my name is Marco and I've said all I have to say."

"Again!"

Marco just looked at the camera, didn't say a word. He was blushing furiously and there was something determined in his eyes. Andrea pressed PAUSE and rewound the tape. *Say it again. Hello, my name is Marco, and I think you're very strange and beautiful.* And then Andrea turned to us and declared solemnly, "I think I've fallen in love!" And Mafalda said, *"¡Qué bien, mi amor!"* and started laughing. "I've met a boy," Andrea said, smiling.

Now, I'm no expert on love and would never claim to be, but there was something alarming about all this. What is love at fifteen? I tried to remember, but my head was throbbing too hard. I felt on the verge of fainting. There was a terrible ruckus in my head, a pain that wouldn't go away. What is love at fifteen? I tried to remember. Love is a large field covered in wild grass and in the middle of the field there is a bush and beneath the bush there is a single tropical flower plucked from the dress of Mafalda and next to the flower there is a single fallen petal and beneath this petal there is a key, but what is this key for?

I don't know, but love at fifteen is finding this key and being amazed and grateful and happy that one has found this key and this key—one is sure, must be the key to happiness. Yes, one holds the key tightly. Meanwhile, flocks of storks fly over rolling hills of red poppies. Waves crash on tropical beaches at sunset. Porpoises frolic. Hands hold hands in dark movie theaters. Yes, at fifteen, falling in love is what it's all about.

Later maybe, one finds that this whole life thing is a long winding corridor through a landscape and this corridor has many doors and the key you found fits in certain doors and doesn't fit in other doors and the doors it opens may not open onto happiness. They might open onto worlds of heartbreak. They might open onto seas of disillusion. Destroyed cities. But when you're fifteen you don't know that. You don't know anything. You don't even know what love is. Or maybe you do. Goddamn, my head hurt.

I looked at Andrea and her bright eyes. She was wide-awake. Perhaps the most awake she'd ever be. "Have I ever been in love?" I asked myself. *I have, but it always ended in cataclysm.* That was my last thought before I passed out again.

The next morning I woke up destroyed. Mafalda was sleeping on the other couch. I went to the bathroom, washed my face. My body was in tremendous pain. It was time to go home. I was a mess. One shoe, swollen face, punch-spilled shirt. All that. I peeked out the front door. Bruno wasn't anywhere to be seen. I was about to leave, but realized I'd forgotten something. I went back to Mafalda and gave her a kiss on the top of her head. Her eyes flicked open and she whispered, "*Adiós,*" smiled, closed her eyes again.

"*Buenos días,* "I heard someone say behind me.

I turned around. It was Antón. He blinked twice, squinted.

"Did you have a good time at the party?"

"Yes, I think I did," I replied.

And then I walked out the front door, and out of nowhere, Bruno pounced on me, took me down. I didn't scream. No, this time I was going to fight for my life. He had me pinned. His fangs were inches from my face. I pushed against him with all my might. But instead of biting my nose off he licked my forehead, gave a toothy smile and walked away. I wiped my face with my sleeve, picked myself up from the ground, and walked out the front gate. My bicycle was nowhere to be seen.

I kicked off my remaining shoe. The asphalt was cool against my bare feet. It was Sunday. I had no idea what time it was. I started walking home. I had only a slight limp. A flock of sparrows flew by. I walked on. After about half an hour, I passed the Apostolic Church of Deliverance. They were singing gospel inside. It made me feel good, all these people singing. The last bits of fog were burning off. I found a little patch of grass wedged between the sidewalk and an empty lot. I think it must have been the only grass for miles. I stopped, inhaled, and smelled the salty tang of the breeze coming off the San Francisco Bay. The sun was already high overhead. I cast no shadow. I was dressed all in blue. Dirty stained blue, but blue nonetheless. Sky blue.

I closed my eyes and rocked back and forth on the grass, felt the planet with my toes. People were singing. We were on a big round rock hurtling through space, and there were people singing.

A fifteen-year-old girl had fallen in love that night. Someone somewhere was riding my bicycle. A cream-colored Oldsmobile rolled by playing hip-hop. Across the street a dewdrop fell from a power line and smashed the wing of a butterfly. Life is cruel and unjust. *I must try and be more truthful*, I thought, *I must try to be a better person.*

Yes.

I let out a little fart and kept on walking.

Reluctant
Damas & Chambelanes

No Quinces, Thanks

BY Barbara
Ferrer

"Ow, Mami, it hurts!"

"*Ay, mija,* calm down, it doesn't hurt."

"Oh yes it does. A lot."

"Pfft. You're exaggerating."

Noooo . . . having my hair scraped back in a half ponytail so tight I could've given Joan Rivers a run for her money in the face-lift department didn't hurt one bit. Nor did having the rest of my very long hair wound up on freakishly huge plastic rollers and covered by a gaudy head kerchief. I'm not talking gaudy in a styl-ish Emilio Pucci sort of way, either. This was just plain ugly. In my denim shorts and bright pink T-shirt, I looked like a miniature Cuban housewife off for a Saturday morning at the grocery.

Miniature because I was all of five years old and already being prepared for a cultural rite of passage. No, I wasn't about to storm the citadel that was Winn Dixie and fight for the one cart without the squeaky wheel, but rather, was about to make my debut as a *dama* in a quinceañera. I'd been invited to be part of the "court" as my cousin's girlfriend made her transition from Nice Cuban Girl to Nice Cuban Lady.

And for this, I dedicated an entire day, beginning with the early-morning visit to the beauty parlor, where I was to have my butt-length, stick-straight mass of hair transformed into bounc-ing ringlets. There's a hole in the ozone with my name on it from all the Aqua Net used to keep my hair in place. Then, there were the lacy white tights and the undershirt and the hoop-skirted petticoat that poofed up to my ears if I sat wrong. Only after all of that, did I finally get to don the stiff, bright red lace and tulle that made up the gown. Who knew it took so much effort to look ladylike?

Finishing the look: Black patent-leather Mary Janes and white gloves that made me feel alternately like Mickey Mouse and Cinderella.

Oh, okay, I will admit the gloves were kind of cool. I mean, I was five and as girly as it got—I spent *hours* admiring them, flex-ing my fingers and holding my hand out for some imaginary Prince Charming to kiss.

All in all, though—the whole thing seemed so . . . excessive. Getting all dressed up for the special quince Mass, having photographs taken afterward with the quinceañera, and the rest of the court—seven couples who were her contemporaries and a "junior court" comprised of seven pairs of young children, all about five and six years of age. I can *still* hear the collective *awwwwww* as we toddled down the aisle, the girls clutching their little white baskets of red, red roses in one hand, the little boys who were our escorts in the other.

Then, after the church pictures, off to my aunt's house so I could "rest" before the big party. I was stripped out of the gown and had my hair reset on the Giant Rollers of Doom to maintain all that bouncy ringlet goodness. A few hours and another ozone layer's worth of Aqua Net later, it was off to the reception hall where the *real* fun started. More formal pictures, the procession, the quinceañera's big entrance, and still *more* pictures. And if I recall correctly, some dance thing with the court. Then, having outlived my usefulness, I was freed from my duties and was able to sit back and watch. Sitting at my table, while my parents mingled and danced, I kicked off my Mary Janes and devoured a bowl of neon orange Cheetos that irrevocably stained the fingers of my beloved white gloves. No one had told me it was a good idea to take them off *before* diving into the finger foods.

Overall, my memories of that night are sketchy at best, but the things I remember, I remember very vividly. Couldn't tell you what the quinceañera's dress looked like, other than it was poofy and white (duh), but I remember my mother wearing a floor-length black-and-white sheath and my father a blue suit. I remember the strange music mix of early disco and classic salsa and the division of generations on the dance floor, depending on what was playing.

Since the theme color was red, I remember an absolute

overabundance of it—the *damas'* gowns, the huge, early-1970's bow ties the male attendants wore with their black tuxes, the flowers, the tulle draping the tables and the stage of the reception hall, the candles set on every conceivable surface that wasn't already covered by food.

And the food. Oh, my God, do I remember the food. Tables groaning under the weight of *croqueticas, sandwiches de jamón, pastelitos de guayaba y queso*, and *cangrejitos* (sweet, glazed pastries filled with *picadillo*) ordered from the neighborhood Cuban bakery. There were cheese platters and *empanadas* and chilled jumbo shrimp. And that was just the finger foods—the *"merienda"* if you will. Because there had to be "real" food, too, you know. *Paella* and *arroz blanco con frijoles negros*, and *arroz con pollo* and *platanos maduros*. As a concession, there was also the fried chicken and French fries, because there needed to be, as the *mamis* sighed, *"comida americana para los niños."*

Then the grand finale, the quinceañera cake, all white, with copious scallops and swirls done in the accent color. (Remember? Fire-engine red. Yeah, every bit as gaudy as it sounds.) You have to understand, too—a celebratory Cuban cake? Nothing like it in the world. Forget the supermarket specials. The cakes you get from a traditional Cuban bakery give a whole new meaning to the word "excess." Layers of sugar syrup–soaked yellow cake, filled with custard and fruit filling, and covered with this singular frosting that's made with so much sugar, you can actually *hear* the granules crunching between your teeth as you chew. And the whole gaudy confection topped with a plastic doll dressed to represent the celebrant. Lord have mercy—if memory serves, these dolls were so ugly, they made the Precious Moments figurines look cute.

Looking back on it now, that particular celebration—my one and only participation in a quince celebration—was genuinely unique. Not just because of the idiosyncrasies of my culture—to me, *those* are normal. But more because I can see it now as not simply a transition for the birthday girl, but a transition for a culture as a whole. Most of the teenagers that night in 1972 were either American-born or had come over from Cuba at such young ages, they had no memories of the country and culture that this celebration represented. Obviously, there were concessions made to that dichotomy—the disco music and the fried chicken alongside the Tito Puente and *arroz con frijoles*. But the heart of the celebration was—and I suspect still remains—Cuban.

In the end, though, what I think I remember most clearly about that long day and evening, outside of the Giant Rollers, was the sense of wonder that I felt. That all this fuss and extravagance— the food, the decorations, the relatives *ooh*ing and *aah*ing (and in the case of some of the *viejitas*, criticizing)—was simply to celebrate one young girl turning fifteen.

I didn't get it.

Fast forward to 1982. Not quite the pastel, Italian-suited *Miami Vice* years and while Christo hadn't yet draped the islands in Biscayne Bay in jillions of yards of hot-pink fabric, Miami was definitely on an upswing. We were still some ways from South Beach being a mecca for anyone other than fixed-income retirees and junkies, but were at least slowly starting to emerge from the shadows of the Liberty City riots and the Mariel boatlift with which the city had entered the decade.

In my little corner of Miami, I was fourteen, and it was just my mother and me, since a few years earlier, my father had done the most clichéd thing imaginable and had himself a midlife crisis that prompted him to leave my mother, clean out their joint savings account, and get himself a girlfriend and a bright red station wagon. Yeah, it was every bit as sad as it sounds.

For him, that is.

Honestly, though, I couldn't be too worried about him and his little red station wagon, since I was too busy being stuck deep in the midst of the World's *Worst* Adolescence. Yes, yes, I know *everyone* feels that way, but how many people can lay claim to the cutest guy in junior high—the guy every girl lusted after because he looked just like a young Rick Springfield as Dr. Noah Drake in *General Hospital*—sauntering up to them in the hallway and claiming, "I'd date you in a second because you've got a hot bod, but we'd have to cut your head off because you're such a dog."

No, I am *so* not making that up. Crass little twerp. Then again, what ninth-grade boy *isn't?* If there's a just God, he's fat and bald now. But unfortunately for my fourteen-year-old self, he wasn't totally wrong. Zits for days and the shiny skin to go with. Straight, limp hair that was so oily, nothing short of dishwashing detergent could control it and, to complete the image, a retainer. My only saving grace from the neck up was that the Coke-bottle-thick, plastic-framed glasses had already been ditched in favor of contacts.

From the neck down I was a classic hourglass—a figure that these days I celebrate. But in the ninth grade, it was just one more thing that set me apart: shoulders and hips for days in a sea of stick-figure hipless wonders. What I wouldn't have given to wear a pair of Sergio Valentes but *noooo* . . . they were designed for the young and the hipless. Nothing may have gotten between Brooke Shields and her Calvins, but my hips got *seriously* in the way of me ever pulling a pair up higher than my thighs.

I'm pretty sure, too, that Crass Boy's disdain wasn't reserved simply for my above-the-neck appearance. Because I was also, horror of horrors, a band geek. *I* didn't find it horrifying—for me, music always was and still is my saving grace. But in terms of Junior High cool points? Not so much. A few might've been salvaged if I'd played something dainty and cute, like the flute or clarinet like most of the other girls in band, but me? Do something mainstream? Nope. I played the French horn and trumpet—both instruments that are historically male-dominated, so on top of everything else, the majority of my friends were guys. In the same-sex, clique-dominated, appearance-obsessed society that was junior high, I was a band geek who hung out with the guys.

Even as desperately as I might have thought I wanted to fit in, there was something in my nature, even then, that kept me from making conventional choices.

Traits that were emphasized, yet again, during that fourteenth year.

It started out with a pretty common scene: me, lounging on the living-room sofa, slowly reading through the Sunday *Miami Herald*, beginning to end, as was my habit, while my mother worked at the table in the adjoining dining room. Somewhere after the Sports page and Edwin Pope's column, I came across this huge article on the increasing extravagance of quinceañera celebrations. This caught my attention because, hey—next birthday, while it was still months away, would be the all-important fifteenth.

"Hey, Mami, there's an article in the *Herald* about quinces and you would not believe the stuff it's saying."

"Like what?"

"Well, for one thing, that people are spending close to ten thousand *dollars* on the parties."

She looked up from the dress design she was transferring from tracing paper onto pattern-maker's cardboard. "*Oye, ¿esos Cubanos están loco o qué?* What can they be spending that kind of money on, anyway?"

Well, ballrooms at the Eden Roc or Fountainbleu, for one, with *nouveau* Cuban cuisine feasts catered. No simple *arroz con pollo* dished up out of the disposable aluminum vats for these parties— nope, now we had a whole *puerco asado* displayed on banana leaves and carved by a dude in a chef's toque, served alongside deep- fried *yuca* chips with the *mojo* poured from silver gravy boats. To paraphrase *The Jeffersons*, we were movin' on up. But not all things changed. The Cuban bakery cakes remained a reassuring staple—every bit as sweet and gaudy and sugar-frosted . . . just bigger.

You could also forget the simple processions from the past. These days, the girls were making their entrances amid swirling lights, announced by the hottest disc jockeys in Miami. They were cruising into the ballrooms in elaborately carved, motorized Cin- derella coaches drawn by Styrofoam steeds or in fancy convert- ibles. Then there was the one that has stuck with me to this day—the girl who was lowered from the ceiling in a huge oyster shell that opened, gradually revealing the celebrant in her gleam- ing white dress as the young pearl of womanhood that she was. Hand to God, I remember it being described like that in the paper—or at the very least, it's the image I was left with.

Sitting there reading that article, I started giggling madly.

"What now?" Mami asked.

"Well, they're describing this one chick being lowered from the ceiling of the ballroom in a big Styrofoam oyster that opens little by little—what if it had gotten stuck in midair, halfway opened? I mean, could you imagine?"

"*Bueno, mija*, it would be memorable at least."

"Especially if she's afraid of heights! Or claustrophobic!" I was practically choking on my Coke, envisioning some dainty chick in a big, white poofball of a dress, desperately trying to claw out of the oyster, bits of Styrofoam shrapnel showering over the court and assembled guests below. Or maybe freeing her by way of the piñata treatment, everyone being given sticks with which to beat at the oyster until it split open and she tumbled out.

Oh, quit thinking I'm so mean. That big poofy dress could serve as a parachute. Or at least cushion the fall.

"So what else does the article say?" Mami asked.

"I'm almost afraid to tell you."

"*¿Por qué?*"

"Because, knowing you, you're going to lose your mind."

You're probably thinking this sounds like a weirdly adult conversation to be having with my mother, right? Maybe it was, but Mami had never been one to sugarcoat or talk down to me, even as a little girl. By the time I was fourteen, while I was still very much her "little girl" in ways that counted, like wearing skirts that were too short and having curfews, she still considered me an intellectual peer.

"You know, I don't get it. These are working-class parents, right? Spending the big bucks, giving their daughters the party of a lifetime—their moment to be a princess."

And, as the article explained, in many Cuban Miami families, when a son was born, a savings account was established for his college fund. When a daughter was born, a savings account was set up for her quince.

"I mean, come *on*." I was getting more and more worked up the more I thought about it. "If I'm reading this right, a boy gets handed the opportunity for four years of higher learning that could lead to a professional career; a girl gets a *party*? Because in the eyes of the culture, she becomes a woman? Are they *serious*?"

"They're Cuban," Mami deadpanned.

"But . . . what about college?"

She didn't bother looking up from her work as she replied,

"What college? Most of these girls are probably going to be expected to get married right out of high school. Or at least be expected to work for a couple of years to save money for when they do get married."

Really, *really* didn't get it. We're talking serious *Twilight Zone* territory, it was that far removed from what I had come to know of male/female roles in my world.

See, I come from a somewhat nontraditional and *very* matriarchal family. One of my great-aunts was one of the first women in Cuba granted a law degree and she then turned around and represented herself in her own divorce. And won. I still remember Tía Nena as this indomitable old lady, gray hair pulled back into a chignon, regal, even in her flowered housecoat. She'd suck on dark *cigarillos* while discussing art and music. She even had theories on world politics that didn't revolve around when Castro was finally going to kick the bucket, either. Talk about revolutionary! Tía Nena's sister, my grandmother, was one of the most well-read individuals I've ever known. To the rest of the neighborhood, she was the crazy old lady with all the cats and dogs, but to me, she was simply Abuela, the only person who could match my passion for reading and storytelling, with towering piles of books in her living room (a habit I inherited and still can't break). With a memory like a steel trap and a tongue that dripped acid, it was no wonder my mother looked worried every time one of the other relatives compared me to my grandmother.

My mother is yet another solid link to this chain. She had graduated high school by fifteen in Cuba, and when she arrived in New York as an adult, she enrolled in night school at an American high school so she could learn the language, get a diploma, and be able to attend a trade school. Then, as a fortysomething divorcée, she opted to enroll in community college and then university to obtain a liberal arts degree, all the while continuing to work as a top-notch production pattern-maker in the fashion industry.

So maybe you can see why the idea of the *boy* getting the college money was a bit . . . mystifying to me. And none of my immediate female relatives, including my older sister, had had a quince. It was an almost foreign concept to me.

Adding to the strangeness of the concept was the fact that despite growing up the child of Cuban émigrés in the Miami of the '70's and '80's, I felt more American. Totally my parents' fault, because they made what was, for Cubans back then, a radical choice: to move to the decidedly *non*-Cuban suburb of North Miami, rather than Hialeah or any of the other Latino-dominated suburbs of Southwest Dade. Their feeling was that assimilation was more important than living in what would undoubtedly be a more familiar, yet ultimately too insular, enclave, with other immigrants and their first-generation children who were growing up speaking English with a slight accent.

Which isn't to say we didn't maintain the culture at home. We did—in a thousand small ways. For example, I was more familiar with *picadillo* than I was with meat loaf and I knew who Celia Cruz was before I knew, say, Dolly Parton or The Carpenters. But while I understood Spanish perfectly and could read it passably, my first language was definitely English and my primary identity was that of a young American girl. Keep in mind, too, that identifying as "Cuban-American" wasn't as commonplace then. You were one or the other. So in my mind and with my parents egging me on, I was totally and completely American. As such, I had all the dreams and expectations that went along with being a young American girl—I was going to go to college and become either a professional musician or a teacher and in a few wild moments, I even entertained thoughts of becoming an Olympic figure skater. (Hey, everyone's allowed a pipe dream, okay?)

To my father's side of the family, many of whom had predictably settled in Hialeah or Little Havana and who spoke nothing but Spanish, even after twenty years in the country, I was, well . . . weird. Or at the very least, vaguely amusing. They'd pat me on the head and say, "Whatever you say, dear," (in Spanish, of course), and shake their heads, convinced that I'd come to my senses and eventually marry a nice Cuban boy.

Oh, please—as *if*.

My future in higher education had never been in doubt. I *had* a college fund collecting interest in the bank. Identified early as academically gifted, I was enrolled in the nationally recognized Gifted Program for the Dade County School System and there was some discussion given to accelerated learning and possibly graduating high school early, even though eventually, I opted to stay and have a more traditional school experience.

I was also musically gifted; my primary instrument the piano. And since my parents were of the type who searched out the very best for their children, I didn't simply take lessons—I was schooled in classical piano by a former prodigy who was renowned through Latin America as a premier interpreter of Cuban composer Ernesto Lecuona's music.

A quince? Not even a blip on the radar. To us, that was something that the *other* Cubans did. The ones who were hanging on to Cuban traditions the way my mother hangs on to a sale item at Macy's, still honestly believing that they had a chance of making it back to the island, someday.

Yeah, I was every bit as snotty and obnoxious as all of this implies. And as life has a way of doing, I was knocked off my lofty pedestal in a *big* way. Remember my father's midlife crisis? How he left my mother and cleaned out the joint savings account? (Okay, not completely—he left us a generous $80.11.) Well, that joint savings account had also contained my college fund and the lovely accumulated interest. (Wonder if that's what he used for the ugly red station wagon.) Anyhow, in one of the rare times I saw him after he and my mother separated, he confessed that he had no regrets over taking the money, because really, I wasn't going to go to college. In fact, I'd probably wind up pregnant at fifteen, so what was the use of holding on to the money?

In other words, he'd merely been paying lip service to the whole Modern Approach to American Parenting thing.

Charming fellow, my paternal unit. Colored my feelings toward Latin men as a whole for a very long time, but that's a whole other story.

So to recap: I was the latest in a long line of Cuban women who were, in one way or another, mavericks. (The polite term for it.) I'd been raised in a decidedly *not* Cuban atmosphere. I was raised with expectations that were definitely not the norm for most Cuban girls my age. I didn't really hang out with many girls my own age, period, let alone, Cuban girls, so it's not like I was surrounded by a peer group for whom a quince was this big thing to look forward to.

I also still had some memories from my one experience as a dama that made me less than eager to pull a repeat performance, except starring me as the celebrant. Now, don't get me wrong. Despite my hanging out primarily with guys and being a big-time band geek, I was still every bit as girly as I'd been at five, experimenting with makeup and loving getting dressed up. However, the idea of going through major league hair and gown torture so people could stare and applaud me for making it to the grand old age of fifteen left me feeling more than a little queasy.

You know, I realize just how bizarre that sounds, how uncomfortable I was with the idea of such intense attention. Especially coming from someone whose big dream was to set the Broadway stages on fire and become the next Barbra Streisand. Back then, I was unable to articulate the difference—these days, I recognize what it is. I have no problem grabbing the spotlight in a role, or as a performer, or to take credit for something I've accomplished. But to be the center of attention, simply *because?*

I'll be over here cowering in the corner, thanks.

Understand, though, if I'd really wanted one, my mother would have moved heaven and earth and quite possibly a few other planets to get me a quince. It wouldn't have been some flashy event at a beachfront hotel with a pink champagne fountain ("because, *mija*, that would be so tacky") and she would've made my gown, which would have been perfectly fine, since the woman is a pattern-maker and seamstress beyond compare.

Truthfully, too, I think she sort of wanted me to have one, if only to say to my father, "Screw you, Luis Ferrer, I can provide for our daughter and give her all the things you would have taken away." I mean, think about it—a scant four years after their separation and not only were we still in the house in which I'd grown up, (which he'd sworn, in Dramatic Cuban Man fashion, that he'd force her to sell) but back to living a relatively comfortable lifestyle. If keeping our house was like a matador waving a red cape at a bull, something like a quince would've been the proverbial sword through the beast's neck.

Have I mentioned yet that the women in my family harbor a somewhat malicious streak and lack subtlety? The celebration might not have been gaudy, but she most certainly would have put a write-up in the *Herald*, English and Spanish-language editions, and made damned sure my father would've received copies. Multiple copies. Maybe even slip a picture in.

But I have to hand it to my mother—for a Cuban mom, she's always been pretty good about not completely butting in and trying to run my life. Of course, she'll tell you what you're doing *wrong*, and she's not above watching *America's Most Wanted* to make certain the latest guy my sister's dating isn't on there, but you know . . . she can't help herself there.

"So, mija . . . have you thought about it?"

I looked up from studying the pictures accompanying the *Herald's* article. "About what?"

"A *quince*."

"Like, do I want one?"

"Yes."

"Mami, are you feeling okay? Not running a fever or anything?"

"That's what I thought. *Sinvergüenza.*"

So yeah, I basically said "no," and she was remarkably cool with that.

"But . . ."

Knew it was too easy.

"What?"

"Are you sure you don't at least want to have pictures taken?"

Ohhhhh-kay then.

This, I understood was actually code for, "I'd really like it if you would take some pictures to at least commemorate this moment—and give me something to taunt your father with."

Mind you, I *loathe* having my picture taken. It's a long-standing thing with me. I've been told more than once over the years that I give off this distinct vibe in photographs along the lines of, "Come any closer and I will cut your heart out." Even as a baby, I looked rather solemn, if not downright disgruntled, in most photographs. Add to that the fact that my adolescence was not going gracefully on the appearance front. Well, let's just say the idea of spending hours being told to "Smile, *mi vida*—just like a super model," was not high up on my list of Things To Do.

Mom knew she was asking a big thing. But I also knew it would mean a lot to her, so I agreed, provided, "I don't have to wear some stupid fluffy dress that makes me look like a deranged meringue." We compromised. She'd make me a short white dress, we'd pick some other dressy outfits and the picture-taking would commence.

This is about the point you expect me to say that I'm really glad I did it, the pictures really captured me at a pivotal moment, poised on the cusp of young womanhood, blah, blah, *blah*. Think again, Kemo Sabe. The picture-taking went well enough, I suppose. We did the obligatory studio portraits, then, wearing my pretty white dress, we went to the obligatory Coral Gables park where I think every quinceañera since the dawn of time has had her picture taken. At least, in Miami. As for the pictures themselves . . .

We're clear on that whole adolescence-not-going-smoothly thing, right? Remember, too, this was 1982, way, *way* before digital photography and Photoshop and all those groovy techniques that would have allowed me to look, you know, human. I mean, soft focus could only do so much. No matter how much Cover Girl pressed powder I slapped on it, my face was so shiny, it could've competed with a reflecting pool.

Then there was the hat. I'm not sure whose idea the hat was—I've blocked it out—but it was this white, wide-brimmed, southern belle number and let's just say that I wear hats about as gracefully as I take pictures. But my mother was happy; happier still when the proofs came back. And then . . . it happened.

I don't know how many of you are familiar with the Big Portrait Phenomena. You know, where studio portraits, usually commemorating some big moment (like a First Communion, wedding or . . . you know, quince) are blown up to movie theater poster proportions. And if that's not enough, then they're encased in an even bigger, elaborately carved, museum-worthy frame—

usually gold—and hung up where everyone who passes through your house can see it. Of course, that's just what mine had. My big mistake was allowing my mother to pick the picture to be blown up for the Big Portrait. Because, wouldn't you know it— with unerring skill, she zeroed in on the one photograph I absolutely hated the most out of the lot. Blown up to 24" x 28" proportions, it was every nightmare I've ever had come true. You can hear my mother now, right? "*Ay, mija*, don't be silly. It's a beautiful picture."

Sure, Mami. What*ever*.

I did entertain a brief moment of hoping that she was planning on sending it to my father. You know, as torture. But that hope was shortlived, when after it was delivered, it was hung up in a place of honor in our dining room where it proceeded to torment me for years. And you know, it tortures me to this day. The first time I took my husband to Miami when we were still dating, he walked into my mother's house and, because of its prime location, spotted the thing almost immediately.

"What a *beautiful* picture," were practically the first words out of his mouth. My mother's immediate response was, "Well, if you ever get married, I'll give it to you."

It's a minor miracle both of them aren't at the bottom of Biscayne Bay. Him for saying such a boneheaded thing and my mother for using words like "marriage" when I'd been dating him for all of six months. But despite the fact that he had a big mouth and crummy timing, we got married and, true to her word, my mother gave us the portrait. The gaudy frame got broken in transit which again, sparked in me a brief hope that I might never have to hang the thing again, but my husband looked at me and said, "Either you pick a frame you like or I'll let your mother pick one."

Yeah, he's mean that way. But my end of the deal was I would choose where to hang it. He agreed, provided it wasn't in his closet. (*Damn.*)

Early on, though, we'd talked about what the picture signified—besides a particularly hideous moment in my adolescence, and I found myself describing the lunacy that is a Cuban quince celebration to my very gringo husband. Being Jewish, he got the whole thing about a ritual or celebration to mark a transition, but as a philosophy major, it was the nuances he was most interested in. And by that time, having come to terms with both my Cuban roots and my American upbringing, I think I was able to articulate it reasonably well for him.

In many ways, it came back to the quince I participated in when I was five. In some ways, so representative of a culture in transition, adapting cherished traditions from an old, loved way of life, while settling, albeit in some cases reluctantly, into a new one. And the reason I think that these celebrations started to become so expansive and elaborate had more to do with the very generous, giving nature of Cubans as opposed to simply an opportunity to show off. We love nothing more than to share our largesse . . . with everyone.

On the other hand . . . let me amend that. In a manner of speaking, it *was* an opportunity to show off, if only because so many of these proud parents had come from Cuba with little more than what they could carry. Many of them had been highly educated professionals, but once in the U.S., they had had to start over, in many cases, in the lowest paying of menial jobs while they earned the credentials to resume their professions or settled into new jobs because the most important thing was to earn money to feed and house their families.

And now they were able to host this lavish celebration for their family and friends. I honestly believe it wasn't simply to honor their child, but also to honor their achievement; an opportunity to raise a glass and toast their success. And it would always be a toast tinged with sadness because every Cuban family I knew growing up carried with them not only the memories of a lost way of life, but memories of those left behind. Loved ones who couldn't be there on the special evening—so the toast would also serve as a bittersweet reminder of how very fortunate we were.

Am I sorry I didn't have a quince? No. Not really. And while the celebration could have also served to honor the achievement of my mother surviving a devastating divorce, with the added benefit of a nice "bite me" to my father, I don't think either of us felt the need. We'd survived with grace and aplomb and what's that old saying? The best revenge is living well? We were doing just fine, thanks, and continue to do so. My grandmother *did* put an announcement of my turning fifteen along with one of the less-hideous pictures in the *Diario Las Américas* and made sure my father got a copy of the announcement. Yep, there's that less-than-subtle streak again.

Of course, as luck would have it, I have a daughter who describes herself as "half sweet, half salsa" and whose nickname around the house is The Diva. Can you guess she has no problem with being the center of attention—for *any* reason? What do you want to make a bet she'll up and decide she wants a quince?

Well, she'd be following in the family tradition of being a maverick at any rate.

The Perpetual Chambelán

BY Michael Jaime Becerra

He has never been afraid of girls. They don't make him awkward or nervous or the least bit shy. Eddie's appreciation of women goes as far back as he can remember. There is the crush on his preschool teacher, on the blond cashier at Lucky's, on a Filipino girl with silver caps on her teeth. There is another crush on an older girl in the townhome complex where he lives. To get her attention he asks for help from an older boy next door. He suggests that Eddie write her a letter, which would be fine if Eddie was old enough to write. The older boy tells him what he thinks she'll want to hear and he scribbles down his suggestions and folds the paper in half. Before going back inside, Eddie leaves the letter on her welcome mat. "Let me take you to dinner," it says. "Let's go to McDonald's."

Eddie was born in 1985. He is the youngest behind two older sisters who are less than a year apart. When he starts kindergarten, they are all together at the same small Catholic school. There he meets another blonde. He's a charmer, even at the age of five. Later in the year, at recess, they are married in a ceremony performed by a mutual friend. Such is his introduction to big, vaguely religious events.

As his oldest sister approaches fifteen, both girls are offered a choice—a quinceañera or a trip sometime in the future—and later, after they graduate college, they will go on a ten-day tour of Europe. So his first exposure to the quinceañera comes with his cousin. He's seven and he wears a tiny tuxedo and a dark green bow tie to match the other chambelanes. He's adorable to everyone there and in the following years he goes from cute to cuter to flat-out handsome, with little of the bony gawkiness that comes with adolescence. It's no surprise that a girl would want him around on her big day.

In 1999 he's fourteen, in the eighth grade, his final year at the same school. Lunchtime is the highlight of his day. At lunch he and C—— sit at the same table, side by side, in the white and navy blue uniforms the nuns make everyone wear. Back when his oldest sister was crowned the Virgin Mary at the school's May mass, she had to wear a long turquoise cape with her white dress. The cape had white fur around the edges and was long enough that they had C—— follow along to keep it from brushing against the floor. Then C—— was in the first grade. Now she's his girlfriend, his first one.

He likes her eyes. They are big and expressive, a deep, deep brown, and when they get the chance to be alone, he holds her close and looks at her looking at him. She's a year older, but when they kiss she must stand on her toes to reach him.

At this time his best friends are a pair of identical twins who also have their first girlfriends. The twins' mother works yard duty at the school during recess and lunch, and she doesn't care for the way her boys cozy up to these girls. She reports Eddie's lunch dates to his mother. Eddie's mother doesn't care for them either.

One afternoon there's a car wash, the graduating class raising money to hold a final dance. Eddie's oldest sister stops by the pizza place where it's being held, and while Eddie and his classmates wash her Tercel, she notices how her brother's girlfriend is always at his side, how she's relentlessly hanging on his arm. Eddie's sister remembers C—— from the May crowning. Now she seems much too clingy and insecure. Eddie's sister comes up with the nickname "Chicle" because this girl's stuck on her brother like a wad of gum. When she tells his other sister about this, they both laugh enough that the nickname remains.

He doesn't tell C—— about it when they talk that night. Most nights he uses the house phone even though he's not supposed to. He waits until his parents have gone to bed and then he calls the movie theater and lets the prerecorded listings cycle through, over and over. He'd prefer to see her in person, but she lives on the other side of town. Eventually he comes to memorize the titles of what's playing—the show times too—but this boredom is worth it, so long as the phone doesn't ring when she calls. The call waiting beeps and he clicks over and they trade the usual eighth-grade gossip. Sometimes they talk for a few minutes, sometimes they talk for hours. It depends when his mother wakes up to check the line. He doesn't live near a pay phone. He doesn't have a cell phone yet.

C——'s birthday isn't until August, but the initial quinceañera preparations are under way. They are a main topic of conversation between them. She asks him to be her *chambelán de honor* and he agrees. Of course he agrees. It's a no-brainer. He thinks being a *chambelán* isn't too different from taking her to a dance, which he's already done. It won't be a big deal.

In late April, on a bright spring Sunday, there's a family barbecue at his house. While the family is outside, enjoying the afternoon and waiting for the carne and pollo to finish grilling, Eddie and C—— are in his room. The story is that they're in the same group for history, that there's a project they're working on, which is why they're alone and why his door's closed, even if it's not supposed to be. He's not about to waste this opportunity. He's rented a movie about cartoon bugs with the thought that it will be way more interesting than anything they might do for school.

When it's time to eat they emerge from the room and the right side of his neck is marked with a hickey—a perfect, purple spot the size of a quarter. It's a bold, territorial marking and its shamelessness sinks whatever chance Chicle has with his mother, his sisters too.

The way Eddie remembers the next part, the first quinceañera rehearsal takes place at C——'s house. His older sister's away at college and his oldest one's working, so he gets a ride from his mother. She drives a green Pontiac and on the way over she reiterates her disapproval of C—— and his being in her quince. The rest of the ride is quiet and the entire time he is determined in his thinking: No matter what his mother says, he will be C——'s *chambelán.*

They arrive and his mother tells him to stay in the car. Her tone is serious, harsh, and it catches him so off-guard that it staggers his resolve. She disappears inside the house and he stays put. He imagines the shock C—— must feel as his mother explains that he can't be in this quinceañera. There is the hope that someone might change her mind, that she'll emerge with a smile and tell him to call her when he's ready to be picked up.

In his mother's version, she finds C——'s mother among the other parents dropping off their children at school. She introduces herself and reveals that Eddie will not be C——'s *chambelán.* She doesn't agree with their being together and there's no hesitation in announcing this openly. She doesn't wait around for a response. Even if she cared what C——'s mother might say, she doesn't have any extra time to argue. In two hours Eddie's parents will land in Las Vegas to spend the weekend celebrating their anniversary.

There is also a third version, one from C—— herself. She

remembers a phone call from Eddie's father. He explains to her mother that Eddie can't be in the quince. He doesn't like his wife's decision, but it's what she wants.

While the exact circumstances are up for debate, the end result is the same. The matter has been settled—he's out. His mother is being a mother, doing what she thinks is best for her son, but this fact doesn't make the moment any less embarrassing for him.

Understandably, things with C—— do not go well after this point. She goes to a public high school, he goes to an all-boys Catholic school, and together they give in to the inevitable. Eddie begins to understand a few things, and these realizations seem truer and truer, more evident and more obvious with the passing of each day. His sisters' collective opinion carries great power. She is not good enough for my brother, his sisters first thought, and hickey or no hickey, he is convinced that their estimation of C—— became truth for his mother: She is not good enough for my son. There will be future instances like the one with C—— and to avoid them he settles on a guarded, defensive strategy. He wants the chance to arrive at his own conclusions before hearing anyone else's. The less they know, the better.

He also retains something else. Without him, C——'s quince won't be the one she wanted. He has let her down. This regret remains long enough to create a sense of obligation to any girl that might ask him to be in their quince in the future. The obligation taps into his deep-seated loyalty. When it comes to quinceañeras, he's able to recognize the pressures the girl is under, the many exposed vulnerabilities and inconveniences, the mortification that often comes with poofy white dresses and tacky white shoes. There is much for these girls to consider—the invitations and the ceremony at the church and the reception after. And everything costs money. The catering and the mariachi. The dress. The tiara. There are also the many layers of scrutiny—her parents, her immediate family, her relatives that she rarely sees, even on holidays, people who only remember her as a baby.

Everything about these quinceañeras is intended to claim: "This is me!" But he knows that for many of these girls this is hardly true. It is how their parents want them to be. These girls can't deny their mothers and fathers, much less hundreds of years of tradition. He is sympathetic. He understands the weight of judgment and he knows what it's like when important judgments don't go your way. If he can help, he will.

A year later Eddie and his father are working off service hours at his old elementary school. They are security guards, walking the parking lot during Friday-night bingo, though half of the time they're sitting in the Pontiac, talking. His father tells him about his past—the time when he worked in the Sol Cerveza brewery, how he came over to California—and Eddie loves these stories, loves imagining his father as a young man, because he loves his father without question. Still, he doesn't always care for the way these stories blend into talk about the present, talk about their family life, about how the older Eddie gets, the less and less he speaks to his mother. All fathers must donate their time to the school, but Eddie's getting paid by another father to work the time for him, even if it seems more like he's getting paid to park himself and listen.

He also makes money selling Twix—one dollar each—to his classmates who are too lazy to walk down to the vending machines. He needs the cash for the weekends, for his dates with L——. They've known each other since the previous summer, but they've only been dating since February, when he heard from a friend of a friend that she liked him. Before Eddie she'd been seeing a guy he played basketball with. This guy had been having problems with L—— and during one of their fights, he handed

Eddie the phone out of exasperation. "Here," he said. "You talk to her."

On the weekends, occasionally after school, he rides his bike to her house, a journey that takes him all through the riverbed, along Rosemead Boulevard, up and down the Montebello hills. He likes her so much he doesn't notice the distance. When she is sick, he makes her a homemade get-well card in the shape of a doctor's bag and rides the twenty miles to deliver it in person.

His sisters like L——. His mother likes her too. She is polite, respectful. There is another barbecue and when she arrives, she does so bearing a homemade cake. When she leaves, Eddie's neck is unmarked.

In conversation around the house he finds himself playing up her good qualities. There is a mature air about her, even at fifteen, and with little effort he can imagine what she'll look like in ten years, though their relationship is such that he has trouble placing himself in any vision of her future. She has a certain chilliness that comes across as indifference. When Eddie thinks of himself with L——, he wonders how long it will last.

In 2000, the summer between his freshman and sophomore years, he is asked to be in his second quinceañera. It's for L——'s friend D——, and this time the circumstances are entirely different. He and L—— will be one of fourteen couples in D——'s court. This time there is no objection from his mother, no discouraging looks as he is driven to the practices or to the mall, where he spends an afternoon looking for a gift with Tinkerbell on it to match the Cinderella theme.

At the Disney store he finds a Tinkerbell watch that other people will buy for D—— too. He is tall, thin from running cross country and riding his bicycle everywhere. He has his hair straightened so he can comb it up in a big, high wave and slick the rest back with pomade. At Mendoza's in East L.A. he rents his tuxedo and his first visit there is the first time his measurements are taken by a tailor. This quince will be another first, the first time Eddie and L—— ever see each other in anything formal. The *damas'* dresses will be pink—Pepto-Bismol pink. He asks if someone spilled a bottle of it when he sees one girl sitting down.

The mass is in East L.A. Afterward the quinceañera and her court are divided into a town car and a black SUV limousine. They stop at a cemetery, where she visits the family mausoleum. The court is posed for pictures, the serious ones that will be given out as *recuerdos*, and then they are off to the reception. From start to finish, everything is videotaped.

They arrive at the Disneyland Hotel, at the Grand Ballroom where serenading mariachis await them. Inside, the room is lavishly decorated. The tables for the quinceañera and her court are distinguished by centerpieces with large crystal carriages, while the many tables for the other guests are adorned with elaborate, homemade decorations. These other centerpieces are tall, wrought-iron structures wrapped in vines and flowers and white lights, a dancing doll on the second tier, a candle burning on top. Wedding food is served for dinner—chicken and vegetables and some kind of potato—and for dessert there's a six-tier cake topped with a castle. Later, to get the dance floor going, the DJ starts spinning salsa and eighth-grade disco and assorted new-wave flashbacks, a combination of "Escándalo" and "Lookout Weekends" and "Just Can't Get Enough." At one point he plays the "Thong Song." Later on Mickey and Minnie make guest appearances, Mickey cutting in on D——'s *chambelán*. Her *chambelán de honor* is a cousin. He's seventeen, over from Guadalajara to escort her for the day. He only speaks Spanish and in the wash of Eddie's memories he will barely be an afterthought.

So far, the night has dragged by. Eddie's been distracted through the dinner and waltz and the cutting of the cake. He has his first taste of vodka, but it doesn't give him any of the giddy anticipation that has seized his friends. For weeks there has been a lot of flirting within D——'s court and talk of a planned sleepover has fueled their practices with romantic possibilities. D——'s mother has reserved rooms in the hotel—a suite for the girls, an adjacent room for a *tío* and *tía* who'll be keeping an eye on them. There's also a separate room for the boys, a room so separate that it's in another building altogether.

His mother and father are among the many other parents in the ballroom, and they aren't in the habit of letting their children spend the night someplace else. Tonight will be a night of fruition for his friends and he is dreading the impending moment when he'll have to leave them to go home.

He asks if he can stay anyway. . . .

A surprise!

They say he can and he experiences that special relief that comes when you're fifteen and allowed to fit in with everyone else. Perhaps it is a lark, his parents giving in to popular convention. Perhaps they have drunk too much, or maybe a father sees all the other sons and feels bad for his own. Perhaps it is because L—— is not C—— and so his parents aren't worried that their boy will return home with the trappings of hanky-panky.

Late in the night, after the DJ has disassembled his spinning lights and after the hotel staff has put away all the tables and chairs, the *chambelanes* sneak out of their room, Eddie in his dress socks, his tuxedo pants, his wrinkled white tuxedo shirt. The air is heady with anticipation and hastily applied cologne, yet he doesn't know what to expect. A few hours ago he didn't think he'd be in the *damas'* suite with L——. Besides, kissing is as far as things have gotten thus far. It's not in him to bully a girl beyond her limits.

At the same time, he's hopeful for something, even if he's not sure what that thing might be. Whenever he's opened up to her, he's left feeling as though he's cast a bottle into the ocean. When he's gone all out, like with the tiny red pager he bought her because it was the one she wanted, his efforts have barely registered a response. After tonight, whatever's been between them will be different.

D—— is still in her big white dress. The other girls are in their big pink ones too. L—— is wearing his tuxedo jacket over her dress and she's quiet, superquiet. Actually all the girls appear nervous. No one wants to get busted, so any boy not already expected by someone is sent back. Eddie and L —— sit in the front half of the suite, on the king-size bed. All the other couples are split between the bed and a fold-out couch, all the couples except for one. They're an established pair and they take the rear half of the suite, the side that's private, for it's known that tonight they will be requiring such privacy.

Eventually the small talk winds down and the couples start coupling up. Eddie inches toward L——. Their improbable moment has arrived and she scoots away. She leaves the bed and walks across the room, to the couch, where another girl, a friend of hers, is going to sleep. L—— joins her and soon enough she is asleep too.

He is disappointed, and not because his hormones have gotten the best of him. Instead he recognizes that this is a night everyone will remember. He looks around the room at the other couples. Tonight will be a tender point of reference for them, a real touchstone, a memory. It hurts that she would want to avoid this, that she would leave without even talking about it, that she would again choose to avoid him in the process. What the fuck? Confused, he waits until the other couples' visits have run their hot courses. He heads downstairs. The early morning is deep blue, the sun about to come up. His disappointment will linger after he's showered and changed out of the rented tuxedo. The feeling will nag on into the next month, through fall and winter. That morning he skips breakfast and catches a ride home with one of D——'s cousins. They speed along the freeway, the sun rising behind them.

His friend A—— is an only child. His birthday is in December and his mother wants to mark his turning fifteen with a mass and a reception. For A—— it's an uncomfortable idea, but his mother insists. She explains the tradition, that boys were originally given fifteenth-birthday celebrations, not the girls. Nevertheless, this is the year 2000, and right now the prospect sucks. She offers a mass as a compromise. He wants to make her happy. He keeps the tradition in mind and his mother's idea begins to take on meaning. If they're going to have the mass, he thinks they might as well have the party.

A—— and Eddie go back, way back to his days of kindergarten weddings and basketball at recess. One day after school he's approached by A——'s mother. In front of the gymnasium she mentions that A——'s going to be having something for his birthday. She asks Eddie if he'd want to be part of it. Eddie's own fifteenth birthday was the usual low-key affair, pizza and ice cream at home with his family. With the mass, the party, A——'s birthday seems to Eddie more like a quinceañera, a quinceañero to be exact.

His experience with C—— has taught him what not being welcomed is like. Since then he's always made an effort, especially when someone's parents are involved. Because A—— doesn't have any brothers or sisters, it's clear that this rite of passage is important. He knows A——'s other friends will give him all kinds of shit, but he won't be the asshole that says no.

There are five guys in A——'s court, four friends and a cousin, all unaccompanied by *damas*. At Friar Tux in San Gabriel they each rent black zoot suits and patent-leather shoes and oversized fedoras. They begin working in A——'s garage on the dance routine which is intended to be the highlight of the party. A——'s mother has hired a choreographer and she's put together a bunch of moves set to a medley of songs she's chosen. They practice on Tuesday and Thursday nights, learning the pattern of steps and counting beats and figuring out how to slide just right. They work on it for six weeks.

A——'s mass is held at their old elementary school. Eddie knows the priest from the years when he was an altar server on Sundays. In his sermon the priest discusses the importance of manhood, how more young men should have such a celebration to welcome them into the adult world.

The reception's in West Covina, in the rental hall at VFW Post 8620. It's old and dark, the walls made of cinder blocks and decorated with framed portraits of soldiers and plaques preserving their memories of war. A——'s *chambelanes* are scattered throughout the room. Eddie sits in the back with his friends, though his parents and his sisters are there too.

The guys look like a boy band as they assemble on the dance floor and "A Little Bit of Mambo" fills the room. There's no denying that it's a cheesy, corny sight, but A—— and Eddie and the four other guys neutralize this cheesiness and corniness by embracing it. The song switches to "Shake Your Bon Bon," to "Diamond Girl," to "Spin It," and the whole time the guys are smiling and walking in stride and doing their best to re-create what they remember from the garage. They spread out, strolling up to the tables of partygoers, snap-snap-snapping their fingers. Their shoes have slippery soles, the kind of slippery that's good for dancing, for making quick, precise turns. They regroup for "Staying Alive," having fun with the hustle, shaking their hips and twirling imaginary six-shooters—the first guy, then the second guy, a quick chain of motion down to A—— at the end.

It all concludes with "My Girl" and for this final portion of the routine, the boys act out the various lyrics in the song. Eddie spins and steps and slides, all Temptations sunshine on a cloudy day, though this part of the routine seems doubly ironic for him since he isn't singing to any girl in particular and L——— isn't there.

They're still together, but he doesn't recall her being with him that night. Because he has stopped talking to his sisters, they have no idea what's going on. They don't know about the emptiness he feels with her, about the rumors that have recently surfaced about L——— flirting with an old friend from elementary school. Eddie's only fifteen, but he knows that love—even if it's love in sopho-more year—should be more than a lopsided affair that is all home-made cards to no one and long bicycle rides to nothing. *Stop treating him like shit*, her friends have told her. *He's going to leave you*, they've said.

In April he does.

Fuck this. Fuck caring. Fuck making yourself available. Fuck trying hard because in the end they won't care anyway.

He gets his driving permit, then his license and a black Nissan handed down from his father. He also gets all the freedom that driving entails. He works in a movie theater after school when he's not running track or hustling after rebounds on the basketball court. He starts in with a different crowd, privileged kids in big houses up in the foothills, the kind of kids who get new BMWs for their birthdays. He starts drinking. He begins with whatever beer he gets his hands on until he rediscovers vodka. There are experi-ments with other things. There's the occasional afternoon where he wakes up to lectures about leaving his keys in the doorknob.

In 2001 he's the *chambelán de honor* in a quinceañera for a dis-tant cousin, a cousin he's seen at Thanksgiving, at other cousins' birthday parties. She's technically a cousin-in-law, and before this he's never had a conversation with her—*ever*. Their rehearsals don't seem very productive. He doesn't really know anybody there and her backyard's so small there isn't enough space to practice everything at once. But it's okay. By this point he's done enough *chambeláning* to not be worried. In the end it will all come together, he thinks. It should be about having fun.

Meanwhile he dates dozens of girls, so many girls that in time his reputation will precede him at other schools. Girls from Ramona and ones from Mission and ones from LaSalle and ones from Saint Paul. He dates through circles of friends and eventually moves beyond the Catholic-school circuit. These girls like him because they think he looks like Morrissey, because he has nice hair and a Smiths T-shirt, because they need a date, because they heard from a friend that he was nice. He meets them in parking lots, in shopping malls, empty movie theaters, their bedrooms when their parents aren't home. He's up-front with them. He tells them from the get-go that he'll only offer them so much.

For the most part, they're fine with this. And as each encounter becomes more and more intense, more and more physical, the girls start meaning less and less. Along the way the word "girlfriend" drops out of his working vocabulary and eventually there are those whose names he can't remember, even though he wants to.

His sisters meet just a few of this assortment. His mother meets two. In fact, his home life's been cut out as completely as possible. His parents are fairly easy to omit—they're asleep when he leaves at night and they're already at work when he comes home in the morning. He cuts out his sisters as well. It's not too difficult to do this with the one that lives in Westwood and then moves to Oakland. But his oldest sister is still in the girls' room upstairs. She worries about him, and when the worry becomes too much, she writes him letters explaining her concern. When he sees these envelopes on his pillow he sighs and wonders what he's done now.

In 2003 he graduates high school. He continues living at home, though he kicks around the idea of leaving, the idea that he and some friends might buy a house, that they'd pay it off together and sell it and split the money. He starts up college in the fall, entering school undeclared. While he maintains his various circles of friends, when he's at school, he's on his own. Now when he drinks, he drinks hard. Usually it's just on the weekends, though there's a period when he goes at it pretty good during the week. This is easy to do. He's filled with a new confidence that comes from being alone and being successful. The cycle of girls hasn't stopped.

He's nineteen, twenty, twenty-one and the offers to escort other girls keep on coming. The quinceañera for his sister's friend's little sister, a girl he's seen but never really met. Another quinceañera for another cousin-in-law. While these events fall through, there are plenty of others to replace them. They aren't very different—a girl's sweet sixteen party, a prom up in the foothills, the gala for that year's Rose Parade. He follows through on each of these, smiling for the photos, even those where you can tell he's exercising that old loyalty to C—— which only exists in his head. In the spring of 2006, a friend asks Eddie if he'd consider taking his cousin to prom. She lives in Orange County, an hour away, but six years before she saw him at a birthday party and hasn't forgotten him since. He doesn't want to be *that* guy, the one prom date old enough to buy beer for everyone else. But if he's her last resort, he'll see what he can do.

He wonders when he'll outgrow these requests, when the rest of the world will realize that he's too old to have a fresh-faced teenage girl on his arm. The following month he moves out with a friend to a two-bedroom apartment in Whittier. He finds himself missing home—his sisters and his parents—sort of, but not really. Maybe it's because he's getting older. Maybe it's because they're getting older. Now when he lets them in on his life, they trust that he knows what he's doing.

One summer night he's downtown with his friends and his sisters and their friends and anyone else who was forwarded the e-mail with directions for the pub crawl. They're among the sleek Los Angeles skyscrapers, and as he's leaving the second of four bars scheduled for the evening, he and his friends run into a young woman on the sidewalk. She's single, a good ten years older, and so far she's been eyeing the other guys on the pub crawl, measuring them up, weighing this evening's options. He introduces himself and the two friends that he's with. Everyone's a little tipsy, though her tipsiness gives her a predatory hunger which can be seen as confidence. She's a real estate attorney, and he tells her he's in real estate too, the sales side, which is true since he's just started showing houses. One friend says he's in school, on his way to becoming a doctor, and the other tells her he works for the gas company. He describes what he does and she skeptically asks about his job title. Eddie leans over to his young doctor friend. Eddie understands what she really wants to know. *Here comes the business card*, he whispers, and sure enough she asks to see it. The friend from the gas company pulls out his wallet and before he can hand her anything, Eddie steps in, getting right to the point:

"Why don't you just ask him how much money he makes?"

There's a burst of laughter, nervous laughter from the woman before she goes trawling elsewhere.

Later that same night, everyone meets at Suehiro's in Little Tokyo. There he orders two rainbow rolls, and after the waitress carries the order to the kitchen, the room begins to spin and everything inside him announces that it will be escaping promptly. His oldest sister helps him to the bathroom, and when they don't return, his other sister goes after them. They make sure the toilet gets flushed, that a mess isn't left on the floor, that his shirt, pants, and shoes are rinsed clean. Back at the table, he lays his head down. His life is a process just like anyone else's. The picture of drunken exhaustion that's taken becomes the wallpaper on his sister's computer screen.

While he doesn't have a handle on some things, he's getting the hang of others. In August he gets a chance with his mother's fiftieth birthday. It falls on a Monday, a work night, so the family gets together on Sunday for the usual birthday dinner: ice cream and pizza and jalapeños on the side. Because his sister calls him at the last minute, he can't be there. He's on a date and all throughout the movie guilt over missing dinner nags at him in a way that it wouldn't have before.

The next day he buys groceries, expensive ones at the gourmet store, and he takes them home, his parents' home, his sisters' home. He makes salad—mixed greens with blue cheese and dried cranberries and candied pecans—and he makes penne with roasted bell peppers and Italian sausage. He makes bruschetta as an appetizer and for dessert he bakes a cake—mango dos leches. It's the first time he's ever used the oven. He follows the recipe, marinating the mango puree in rum, soaking the cake with sweetened milk. He prepares everything without any help and he's made enough of a meal that it won't fit on the table all at once. When his family sits down to eat, he's content to watch.

The Dress Was Way Too Itchy

**NOTRE DAME
HIGH SCHOOL**

SAN JOSE, CALIFORNIA

1974-1975

STUDENT BODY CARD

Monica Palacios
Name
2061 Cinderella Lane
Address

BY Monica
Palacios

I felt bad telling Tony I couldn't be in her quinceañera. She was a good friend, but me? In a dress? In *that* dress? Hell no! It was a huge hoop skirt—you know, Little Bo Peep, puffy this, puffy that—with tons of lace and itchy fabric. What *do* you call that—fiberglass?

Antoñia wasn't my best friend but she was cool. She didn't let everybody call her Tony—so it was a great honor. We were freshmen attending Notre Dame, a private all-girls Catholic high school in downtown San Jose and there weren't many Chicanas in the entire student body, so I wanted to be all "*si se puede*" and do it for "*la raƷa*," but dang—that dress made me nauseous.

"I can't be in your quinceañera because I'm allergic to the dress." That's how I broke it to Tony. I know it sounded lame but I had to get out of it and I figured if I went with a bodily ailment, she would understand. "It makes you sneeze?" she asked, confused, like Gladys Kravitz from *Bewitched* when Samantha would try to explain why there was an elephant in the kitchen.

"No, it's too itchy. It irritates my skin and makes it all red and I have to scratch it. Then it bleeds."

"Wow, you're really sensitive." She felt sorry for me.

"Yeah. I get sunstroke and carsick too." I thought I'd throw that in to make me sound real pathetic. "But if you need help with anything like making the *mole* sauce or creating a hairdo for you, just let me know."

She shook her head and kind of smiled but I knew she was disappointed. Maybe she assumed us California girls had these ceremonies all the time like they did in Dallas, where Tony and her family had moved from a year ago. I wished she had given me credit for at least trying that poor excuse of a dress on.

A few days later, I even shocked my mother when I told her I got as far as trying on the dress. "This way Tony will see that I was making an effort. That I care."

"*¿Pa' qué?* You don't care," she said, lying back with her head in the bathroom sink, eyes closed so the dye that I was applying to her hair wouldn't sting. The shade was Chestnut by Nice 'n' Easy, the color favored by all working-class Chicanas circa 1974.

"How come you never made any of your daughters have a quince?" I asked, massaging her scalp with my Playtex rubber gloves trying to be all salonlike. I pictured myself in a pink smock with a nametag that read COOKIE.

"They're ridiculous," my mother said. "Why give a fifteen-year-old a wedding?"

"Yeah, you're right," I said. "I would have asked you guys for the chunk of money."

"*Qué chunk y qué nada*," she said.

No sooner had she squelched my financial request then she jumped out of the chair. "*Ay*, the beans need water!" She ran into the kitchen unaware of the dye dripping down her forehead. "I told you to keep an eye on the beans! They're burning!"

I watched her douse the beans with water.

Yeah, I should have paid more attention but she was right, I didn't care.

That day when I had gone to try on the dress, I made my friend Lola go with me, thinking she would be sympathetic. "You know how it looks wrong when guys put on women's clothes?" Lola asked.

"Yeah?" I said, looking in the mirror, checking out my front, my side, my ass, trying to figure out if I looked like a baked potato or my Tía Cuca.

"Well this"—she pointed at my reflection—"is the same thing. The exact same thing."

It was true. I looked like a drag queen but she could have been nicer about it. "Lola, did you ever think you could be hurting my feelings?"

"Yeah, I guess—but I'm right, right?" she said, stretching "right" into two syllables.

I nodded my head yes.

"And you were never going to be in this *pachanga* in the first place."

I started scratching my neck. "I know, but I'm trying to be gracious."

"Gracious?! You never use words like that when you're talking about me."

There was this jealously thing that Lola had with Tony which was dumb because I was closer to Lola than I was to Tony. Lola and I went way back to kindergarten where she actually socked Joe Vanotti, the only "Eyetalian" (as my dad would say) in our class, really hard in the stomach because he grabbed a bag of M&M's out of my hand. I knew right there, in the middle of the playground at Mayfair Elementary, standing with the M&M's confetti all around us, we'd be friends *por vida*. Then this nasty stomach itch brought me back to reality.

"Don't just stand there, help me get this sorry-ass thing off me! My *panza*'s on fire! Hurry, hurry! The zipper!"

"Okay, hold still." She struggled with the zipper, making all these weird faces.

"Fuck, Lola, help me! I'm dying!"

"Shut up, *tonta*! And stop moving!"

Finally the dress popped open and I jumped out of it, scratching my body all over the place. "Ahhh, *ay*, sheesh, oh, oh, aaaahhhhhhh . . ."

"Dang, Monica, the dressing-room clerk is going to think we're having sex in here."

Even though I looked like a contortionist scratching my back and my leg, I still managed to get out a, "You wish."

She smacked my shoulder. "At least I'm not a Catholic-school slut," she said, walking out of the dressing room.

I stuck my head out the door. "You're the one who's the public-school *puta*." She shot me a glare and we both laughed.

Lola and I were always teasing each other about our different schools. My parents pulled me out of public school after the fourth grade because my teacher Mr. Hammer strongly recommended a private institution for a smart kid like me. Lola's family couldn't afford the switch to Saint Patrick's Elementary, which then led to Notre Dame. You would have thought this would have torn us apart. But it somehow made us closer.

A *whole* three weeks after delivering the bad news to Tony, I still tried to avoid her at school as best I could. But that was stupid given our campus was as big as a shoebox and because we always managed to change into our gym clothes for PE at the same time.

"So how's everything going?" I tried to be a cheerful fibber as I unbuttoned my white uniform blouse.

"Mmm, okay—I guess. My dad and mom are fighting whether to cap the budget at eighteen thousand."

"That's a lot of tortillas." I was playing it off like it was no big deal but I kept thinking, *I could buy a Porsche for that amount and have some change left over*.

Tony stayed quiet, changing into her white shorts and shirt in front of her locker. She sat on the bench and put on her tennis shoes.

"Did you just not want to be in it?" She tied her left sneaker and then looked at me straight in the eye. " 'Cause if that's what you—"

"No, I mean yes. I wanted to be in it, of course I did. You know you can count on me, but that dress was going to be impossible." I showed her my forearm, "Look at the rash I still have." She stretched out her neck trying to catch a glimpse of my dress disease. "See how red it is?" Luckily my cat Snowflake had scratched me that morning before I left for school.

"It looks like a scratch," she said suspiciously. "From a cat."

"Well that's the drag, man. This rash looks like so many things. I wish it *was* just a simple cat scratch. Those don't itch." I was definitely going for an Oscar. "But this stuff"—I waved my fingers over Snowflake's mark—"this is evil. I'm trying with all my might right now not to scratch it." She bought it. I knew if I brought it back to the itchy, she would feel for me.

"Yeah, sorry about that. Marcía also was saying the dress was bothering her but she's sticking it out. Do you think the dress is too old-fashioned?"

"Do you think it is?" My journalism teacher had told me, to get out of a sticky situation in an interview, always throw the question back at them.

"Well, kinda, yeah, my mother loved it and really, I'm doing this whole thing for her. If it was up to me, I'd have a party without the big hoopla and go to Hawaii."

I felt better knowing she wasn't that into it. And I guess I could have told her I hated the dress but I was trying to be polite, you know, *gracious*.

Now, I knew quinceañeras existed; it just never entered my mind to have one. These crazy-ass rituals were for girls who were born in Mexico, not Chicanas like me. You know, Old World customs that promoted virgins, big hair, and big cake. That virgin thing was so stupid. "She has to be a virgin in God's eyes so he can bless her and send her out into the world." I heard one of my mom's friends say this once at the little café that was part of our church where these ladies cooked and sold *menudo* every other Sunday. What really pissed me off was that this same *señora* then said, "Mrs. Sanchez didn't allow her daughter to have a quinceañera *porque* she lost her . . . ," the woman tried to whisper as she emptied a can of hominy into the huge pot. ". . . virginity." Yuck. I remember I gave that woman a dirty look and walked away.

It wasn't fair. Guys fucked all the time and they were treated like kings. The minute a girl had sex with *one dude*, she was a *puta*. At least the girls I knew were on the pill. They'd all walk together the four blocks from school to Planned Parenthood. Ah yes, the perks of attending a Catholic high school. That whole religious ceremony just reeked of old-fashioned crap. I was about being mod, the hip chick of my friends. And I was always making some kind of fashion statement in my subtle tomboy way. Like the time I took my 501s, slit the sides of the legs, sewing in some hippie flower pattern, creating my own super bellbottoms—jealous? Yeah, I was a trendsetter, so it was just a matter of time before I cut my long thick Mexican-girl hair into a shag. I went to Macy's because I had to do it right. David, the snappy hip hairstylist with the perm asked, "Do you want the Fonda?" I knew he was referring to Jane's cut from *Klute*—I was onto his cool cat ways. But I certainly didn't appreciate him cutting my hair while holding a lit cigarette between his fingers the entire time—freak!

The abundance of layers gave my head more curls, and I liked that. But when I showed up for basketball practice that Monday my teammates took one good look at me and announced: "Don't worry, it will grow back quick." "You guys are jerks," I said, you have no sense of style." Spinning the ball on my index finger, I walked away from my pack of varsity players. I don't mean to brag, but I was the only freshman on the team.

Trotting to the other end of the court bouncing the ball, I noticed Tony was also in the gym with the other cheerleaders. They were rehearsing for the big game in two days. I appreciated their enthusiasm for it all, but there was only so much they could do with a team name like The Gremlins. Although I did like watching them do those splits and shit. She came over to me. "Don't let them bother you. I like your hair." She smiled looking at my 'do, as if I had a cat sitting on my head dressed like Peter Pan. "It's cute." She raised her hand kind of touching my strands with her fingertips. "Thanks. I wanted to get it cut now so it would be a little grown out for your bash." That wasn't the real reason, but I was reaching out and you know, kissing ass.

Then Coach Dixon yelled out, "Palacios!" and she sounded like the sergeant from Gomer Pyle. I had to get my butt back to the court and fast. I shrugged my shoulders at Tony, she nodded her head, and I bounced back over to the jocks.

The following morning was Saturday and I usually helped my mom with the ironing. As I was placing the crisp shirt she had just pressed on a hanger, I asked her why she hadn't taken Tony's mom up on her lunch offer. They had finally met in the fall at our Mother-Daughter Freshman Harvest Lunch. "She always asks me, 'When's your mama coming to my house for lunch?' And I usually say something dumb like, 'When she learns how to drive.' So, when are you going to call her, Mom? That was six months ago."

I could tell by her expressionless face and the way she was tightly gripping the iron that she would have rather been talking about enchiladas. "Mrs. Esparza tries to be younger than she really is." She handed me the shirt and quickly got another one.

"Tries to be younger?"

"She has to be forty-five or so and she acts like she's thirty."

"She always seems to act her age when I see her. What did she do that screamed thirty?"

"Her skirt was short and she had on too much makeup."

"Mom, at that bogus luncheon, she was wearing a pant suit. You even said it looked like your blue one, remember?"

"Oh, yeah. But she had on too much makeup. She looked like a *chola*."

I stared at her for a while. "And that's why you're not going to have lunch with her?!"

She stayed quiet as she continued creasing my dad's pants on an ironing board she had since the 1950's. I wanted to open up my mom's brain and find out what she was really thinking. She was in no way encouraging when I started telling her about Tony's ceremony months ago. "So you never wanted me to be in Tony's quinceañera in the first place because you thought *I* was going to turn into a *chola*?" I think she believed Mrs. Esparza was flashing her *lower*-middle-class status while we quietly existed in our *upper*-lower-class domain. "Do you think Mrs. Esparza is being too showy by giving her daughter this expensive *fiesta*? Does it bother you that we can't afford a party like this?"

"*Como qué* we can't afford it?!" I knew this subject was going to piss her off. "We can afford it. Your father and me don't think it's necessary to throw all that money down the drain. Then she changed the subject. "Get me another *cosa* of starch."

I got up and went to the hall closet to get her another can of Niagara Starch. I was mad, but I couldn't do much about it.

"Mom, just so you know, *cholas love* to use starch. So technically, you're a *chola*, Mrs. Palacios."

"*Ay, qué la* . . . Get out of here." She punched the air with the iron—this chick was fierce. "Go do your homework!"

"It's only nine in the morning," I protested.

"Then go make me some pancakes!"

Before we knew it, the school year ended and the hot days of July had begun. Time flies when you're avoiding being Catholic. The quinceañera month had finally arrived. The party was set for Saturday the fourteenth, and the weekend before, my parents had decided to go out of town. So Lola and I watched *The Graduate* at my house until two in the morning, enjoying our own version of the good life: Hawaiian Punch and vodka. I had seen this film three times already and thought Mrs. Robinson was foxy. Lola didn't think she was so great.

"That's messed up she sleeps with her daughter's boyfriend. If my mom did that, I would disown her." She took a big swig.

"If my mom was Mrs. Robinson, I would marry her."

"That's sick, man."

"I just mean . . . she's really beautiful. 'Cause, like, if my mom was a movie star—I wouldn't be dealing with no quinceañera party."

"Ah man, do you always have to talk about that? You're not gonna be in it after all, so why do you keep talking about it? And . . . why are you even friends with her?" I knew this question was coming.

"She's my school friend, man. You're my *friend* friend. You and me go way back, chick. I have a different relationship with you."

"She acts white. She tries to be all good and shit." Lola said shaking her head around like she meant business.

"Acts white?! She's having a quince and she speaks Spanish to her parents! Doesn't seem white to me. I don't see *you* speaking Spanish to your folks."

Lola stuck her red tongue, at me.

"Real mature, dorko. Do you think I act white?" I asked. I didn't think I did.

She shrugged her shoulders. "I don't know . . . maybe . . . sometimes. I mean like, since you been going to Notre Dame, you seem a little different." She slurped her rocket fuel.

"I'm acting white because I go to a private school?! That's lame, man. You're sounding like those three stupid *cholitas* in my class who always ask me: 'Hey, Palacios, how come you don't hang with us, *esa*? Are we too Mesican for you? How come you have white friends?' They don't even say Mexican right."

"Why *don't* you hang out with them?"

" 'Cause they're punks. They look like thugs and all they do is try to act all bad and harass the other girls. That's not my style."

"Why *do* you have white friends?" she asked defensively.

"Lola, I can't believe you're asking me this. You had a white boyfriend. I've never had a white boyfriend."

"You've never had a boyfriend!"

"Fuck you, I've had boyfriends!" I grabbed a pillow from the couch and threw it at her. She ducked and her movement made her spill her cocktail on her Mickey Mouse shirt.

"*¡Cabrona!* My favorite T-shirt!"

"Stop crying and take it off." Lola whipped off her shirt to discover two Cheetos stuck to the tiny little flower in the middle of her bra. We were both surprised but started cracking up. Every now and then Lola and I would get into these arguments and get pissed at each other and then in a matter of minutes we would start laughing hysterically. I never laughed like this with Tony.

There was no getting around it; I had to attend the function. And even though my mom would have preferred that I not RSVP because: "You're too young . . . there's going to be booze and *muchachos feos!*", she had to let me go. Although Lola made a big deal about how she wasn't going to go with me, she went. Yeah, we kept saying we didn't want to go, but this was going to be our first quinceañera and now that the day had arrived, we were excited.

My mom made it clear that we couldn't go by ourselves, so my two older sisters Eleanor and Marty said they'd chaperone us. But the real deal was, they were going to drop us off and go on a double date with their good-for-nothing boyfriends, or as my mom called them, *greñudos*.

"Wear a dress. Look like a lady for once." My mom tried to persuade me to go against my policies but she failed—once more. Wearing pants was very natural for me, even as a little kid. And yet again, I managed to put together a cool look: my sexy shag, hiphugger superbell-bottoms—purple with thick brown vertical stripes—huge platforms, and a purple blouse that was shiny and had ruffles down the middle and on the cuffs. Marty said I looked like I was in the band Paul Revere & The Raiders. And I had this really cool brown belt with a thick buckle I just bought at Mervyn's with my babysitting money. I had really wanted to impress the crowd by wearing a tuxedo but I knew people would freak and mistake me for a *chambelán*. Lola, on the other hand, looked like Cher with her slender figure and eyeliner electricity. She went wearing a black miniskirt, a black leather halter top and these bad-ass black suede go-go boots. Come to think of it, I guess I did look like Sonny. Regardless, we were styling.

We didn't go to the church part of the event. I was always going to church because of school and Mass on Sundays with my parents, so the thought of going to Mass when I didn't have to was just plain wrong. We went directly to the reception at the lovely Azteca Hall, one of San Jose's finer venues. I think it was next to a prison. I made Eleanor drop us off at the Der Weinerschnitzel across the street. My plan was to have a corn dog and a Coke—get something in my stomach before we started drinking. And I didn't want to chance it by eating the *mole* at the reception because the last time I had *mole*, the chicken was all rubbery and it made me sick. Lola and I sat in their outdoor picnic table area, in our fine threads, checking out the crowd arriving as we crunched on our dogs.

We watched this long-haired *cholo* drive up in a candy apple red lowrider Impala with a chain steering wheel. He parked, got out showing off his white tux, went over to the passenger side, opened the door, and helped Tony out of his fancy ride.

"Cool, a lowrider." I was shocked, but I gotta say, it was impressive.

"Isn't that her cousin?" Lola asked as she chewed on the corn dog stick. We had checked out a party once by the San Jose Flea Market where we met this dude and Tony's other cousins.

"Yeah, Pepe." I didn't think her parents would allow her to cruise in a lowrider. Tony's entire court, seven chicks and seven dudes, were also getting out of their cars by that point.

"Hey, Lola, see that chick with the shortest hair?"

"Yeah," Lola said.

"Abortion. Tony don't know."

She gave me a concerned look as she sipped on her large Coke through a straw.

"Pancha spilled the beans. Don't tell no one," I said firmly.

"Yeah, like I'm gonna run across the street and yell, 'Baby killer!'"

I let out a sigh and stared at her as if I were saying: Don't be stupid. "I'm just telling you, Lola. If Tony found out—Pancha would not be looking all happy and wearing that dress."

"Man, who cares? Pancha made the right decision for her. People need to open their minds," she said, trying to be Miss Profoundness.

"You're trying to be all adult and you have mustard on your nostril—what a burn!"

"No, I don't," she threw out, hitting the table lightly with both fists.

I pulled out my compact mirror from my hippie leather bag and shoved it in her face. "Check it out, *esa*."

Lola grabbed the mirror from my hands and sure enough she found mustard on her nose. She started laughing, which made me laugh. Then she got up and started dancing with the mustard on her face. "Yeah, I would have been all bad and shit and dancing with this mustard on my schnozz." She started dancing all spastic-like and that made me laugh even more.

"Lola," I gasped, "stop, please!" I was wheezing now and

making high-pitched noises that only dogs could hear. Then I felt a squirt of pee escape my body. "You're making me pee. Stop it!"

"Then shove a Kotex down your pants!" She kind of sang the line and let out a James Brown squeal. And she kept repeating the line and the squeal as she created the dance *Shove a Kotex Down Your Pants*, pretending to slide a sanitary pad "into her pants." By this time I was crying, she was just being so damn funny. I doubled over holding my stomach, praying that if I tightened my *nalgas* together, I could hold in my pee. And it worked.

As I sat there catching my breath, wiping away tears with my corn dog–stained napkin and making sure I wasn't sitting in a puddle of urine, I looked across the street and thought, *That could've been me in that huge baby-yellow, itchy-ass dress*. But everybody seemed happy being there in that parking lot, with their partners checking each other's attire, making sure nothing was out of place. How nice for them—glad it wasn't me. And as soon as Lola came out of the restroom we made our way over there.

We walked into the joint and the ten-piece mariachi band started playing my favorite mariachi jam, *"La Negra."* The place was packed and there were kids sliding past us on the floor. It was like a Mexican Woodstock: old, young, war veterans, women giving birth in the back, the pretty and the hip—that would be me and Lola—lots of sideburns, lots of wiglets, buckets of fake eyelashes. And there were so many different scents coming at me: Old Spice, chorizo, Aqua Net, Pine-Sol, Jean Naté, Tide detergent, tequila—I didn't see any Jose Cuervo bottles but I sure as hell smelled it.

Before we knew it, the mariachi madness turned into a waltz, and Tony and her father walked to the center of the dance floor. This sent everyone to their seats except her *damas* and *chambelanes*, they remained standing and off to the side. The room fell silent.

"Dang, it's so quiet," Lola whispered.

"I know, huh? Getting all this *raza* to shut up—miracle."

We stood mesmerized like the others, watching Tony and her dad start the first dance. This part of the quinceañera is probably considered the most symbolic gesture in the ceremony because he is giving his daughter, a woman now, to society. Another dumb tradition, I thought. Mr. Esparza tried to be all stoic but he ended up sobbing, probably realizing how much this whole Mexican bar mitzvah was costing him.

Tony looked beautiful—radiant in fact, in her white *wedding* gown. She kept reporting to everyone in homeroom that she was doing this event for her parents but *Miss Esparza* was certainly enjoying the moment—she liked being the center of attention. She was good at it. She was class president, a straight-A student, and a cheerleader. She had represented our school for the Christian Youth Organization Spring Conference. Tony was a good girl and good friend but she didn't have fire in her eyes like Lola *la chingona* standing next to me, arms folded, hip out, head tilted back with her wannabe *chola* attitude. But Tony was enjoying this dance with her father and that's what mattered. There was a huge applause and her court sprinkled onto the dance floor. The boys gingerly leading the girls: left hand to right hand, held up at eye level as they walked side by side. It was nice and all but I just kept thinking, *You would never see this on* Soul Train.

As everyone took pictures and admired the ballroom dancing, Lola asked me, "Hey, how come they're waltzing and not doing the cha-cha?"

"It goes all the way back to the Spaniards conquering the Aztecs and making them do shit they did in Spain. Know why she's wearing a crown?" I thought I'd continue to impress her with my brilliant knowledge. Yeah, I was an intellectual—I owned a library card.

Lola shook her head no.

"She's a queen before God."

"What a burn," Lola let out. We looked at each other and said at the same time: "Wish I was stoned." That made us bust up laughing and the *viejita* next to us gave us a dirty look so we moved to another spot.

"How come you know about this stuff?" Lola enquired as she stared at this cute guy across the dance floor who was checking her out from head to toe.

"Tony was telling a bunch of us a few weeks ago at lunch."

The cute dude then invited Lola to go outside and smoke a joint; all through sign language.

"Hey, that foxy guy just—"

"I know, I saw him. Are you going to leave me this early?" I asked. Lola was always getting hit on when we went out. She was just like the song: tall and tan and young and lovely.

"No, chick. But let's meet up with him later, okay?"

"All right," I said, and checked out Lola tell the guy with her hands and eyes, we'd catch up with him later. He agreed. It was like watching a sexy silent movie with close-ups and everything.

As soon as the mariachi band ended their waltz, the very loud band on the other side of the hall started up with Santana's "*Oye Como Va*" and the crowd went wild, swarming the dance floor. The musicians, mostly Chicanos with one black dude, were called Life and Salvation. It said it on the bass drum and the words vibrated as the drummer pounded away. These guys had that Santana look as well: lots of hair worship, you know, beards, mustaches, goatees, long hair, afros. They were wearing leather, dirty denim, lots of leather wristbands, motorcycle boots and ridiculous platforms. And I remember all of them had dark sexy eyes. Yeah, they looked dangerous and we were digging it.

Lola and I danced our way onto the floor, screaming *oye como va* every time the band sang it and people stared at us —annoyed.

"They think we're drunk!" I yelled to Lola.

"Skunk?!" Lola yelled back.

"No man, they," I said, pointing to the people, "think we're drunk!"

Still dancing, Lola held up her hands and shrugged her shoulders and shook her head no. I went right up to her and yelled into her ear canal, "THEY THINK WE'RE DRUNK!" She pushed me away and screamed out, *"¡CABRONA!"* Everybody gave us dirty looks on that one, so we started moving around the dance floor. As we traveled we eventually ran into Tony and her group. I caught her eye, she smiled at me, and I walked over to her while Lola went the opposite way.

I stood in front of her and gave her the biggest smile as she danced with her cousin. Tony stopped and hugged me, swallowing me up in her puffy prickly dress. *This is what it must feel like to make out with a porcupine*, I thought. Fabric dust found its way into my mouth along with her cloud of Chanel No. 5 perfume. I politely turned my head and coughed and then sucked in a glob of clean air. "I'm so glad you came," she shot into my ear. We let go of our embrace and I mouthed to her: *Thanks for inviting me.* I turned to her cousin Pepe and gave him a macho *movimiento cholo* nod, he reciprocated—he was cool. I waved good-bye and weaved and bobbed my way through the crowd, searching for Lola, who was standing next to the champagne fountain holding up her plastic glass, motioning me to come and quench my thirst. I floated over with my smooth dance moves and she handed me a full glass.

"Drink up, Monica, because today is your lucky night, baby. We are going to par-tay."

With that, we tapped our glasses, gulped down their entire contents, and the evening became more interesting.

As we stood practically under the fountain (this way our glasses were always full), Mrs. Esparza sailed by. "Monica, thanks for coming, *mija*." I liked when she called me *mija*. I quickly handed Lola my glass and gave Mrs. Esparza a respectable peck on her cheek. I was getting ready for her to scold me because I was drinking but instead she said, "*Mija*, make sure you eat enough so the champagne won't make you sick. Go get a plate of *comida*." She pointed to the buffet tables.

"We're fine, Mrs. Esparza. This"—I pulled Lola over by her elbow as she held our glasses—"is my friend Lola Sandoval."

"*Mucho gusto, mija.*" Mrs. Esparza greeted Lola and looked down at her two glasses, looked at me, back at Lola, and then back at me. "*Mijas*, go get some food." Then some woman in this hideous blue glittery dress that was way too tight for her started talking to Mrs. Esparza and they both melted away into the crowd. Once Tony's mom was gone, Lola stood there with a smile and said, "*Gracias*, Mrs. Esparza. It was a pleasure meeting you. This is a lovely party. Great color scheme."

"Enough with the clever compliments, Eddie Haskell. She's gone, *esa*."

"Whew, that was close and really cool of her," said Lola. "My mom would have already thrown her *chancla* at me."

"That's why we have to maintain being politely drunk—okay? No taking off the halter top in the middle of the dance floor."

"That's gonna be hard, but I think I can manage."

"Let's go get a plate of food just for show and then we'll ditch it and continue our liquid diet," I said, being the boss of her—as usual.

* * *

We got our plate of food, walked over to where Mrs. Esparza could see us, and pretended to eat. We made eye contact and then casually walked out of her view and handed our plates to the army of apron-wearing chubby little *tías* who were maintaining the buffet tables. We wasted no time. We each got one of those sixteen-ounce plastic Budweiser beer glasses and filled those suckers up with more champagne. Then the band started playing the instrumental version of "Grazing in the Grass" by Hugh Masakela, which was always a *raza* pleaser, and we couldn't help but find ourselves on the dance floor with the masses again.

As we were doing our thing in the middle of this *mexicano* watusi moment, it dawned on me that a quinceañera was really an inauguration into Chicano family partying. I mean, here we were getting tipsy and dancing to hip music amongst *tías y tíos, abuelitas,* moms and dads, hard-core *cholos* . . . I even saw the priest doing The Twist. It was a divine spectrum of brown pride—and I loved that I was part of it.

Life and Salvation continued to play my favorite tunes: "Who's That Lady," "*Sauvecito,*" "Come and Get Your Love," "Evil Ways," and "Sex Machine"—to name but a few. Lola and I danced and drank for about an hour straight and then I started bumping into people, which confirmed I was officially drunk. As best I could, I made my way through the crowd and headed outside. Before exiting, I looked back and saw Lola still shaking her money maker.

I walked into the parking lot and found a cute VW Bug to lean on. The cool air felt good on my face and in my lungs. About five cars away I spotted a group of men standing around drinking and smoking cigarettes, but I could smell weed, too. Though you think you can, you never cover up the smell of grass. One guy dressed

all in black shouted, "Hey, *mami,* come over here and share a doo-
bie with us." Man, that was the last thing I wanted. "No thanks,
guys. I'm fine," I shouted back. The dude in black said something
to his buddies then made his way over to me. "Hey, pretty fine
thing, my name is Victor." He gave me his hand. I raised mine and
waved.

"Let's go somewheres you and me and smoke this." He pulled
out a joint from his breast pocket. "This is really good *mota,*
baby."

"Don't want any right now, thanks."

I started moving away from him, feeling nauseous.

"Hey, I'm not through with you, *chavalita.*" He grabbed my
upper arms and pulled me real close to him. I could feel his hard-
ness through his pants. Maybe because I was drunk, I didn't feel
scared. "C'mon, baby, let's go party. You know you want to."

He had me pretty tight in his grip, but that didn't intimidate me
so I told him, "Listen, Victor, I don't appreciate your little dick
pressing against my leg and if you don't leave right this minute,
I'm going to puke on your pretty suit."

He let me go in a flash and I turned and vomited in a nice little
puddle next to the tire of the Bug. I could hear the other guys
laugh at Victor—mission accomplished. Miraculously, I didn't get
any on me—practice. Feeling a whole lot better, I headed inside to
the restroom to splash water on my face and look for Lola. Checking my watch, I saw that it was already 10:30 and my sisters were
going to come for us at 11.

I searched every corner of Azteca Hall and no Lola. I went outside and looked everywhere and asked everyone including stupid
Victor and his friends. They didn't scare me, but what freaked me
out was the thought that maybe they had Lola in the trunk. They
said they didn't but to make sure I meant business I told them, "My

tío Ramiro, Mr. Esparza, has connections to the mob and if you guys are lying to me, I swear, he and I will come after you."

"He's your *tío*?" Victor asked, genuinely concerned.

"Yes, so don't fuck with me." I walked away. My acting was getting really good.

As *I* was about to go into the hall again, my sisters pulled up. Shit. I walked over to their car. "Um, I'm looking for Lola. I can't find her. Did you guys have fun tonight?

Eleanor wasn't amused. "I'm going to park and come inside and help you find her. Is she drunk?" I hesitantly nodded my head yes. "Are you?" she asked, pointing her finger in my face.

"Not anymore," I said, trying to score points, but I think I lost about one thousand.

"Monica, you are going to be in so much trouble when Mom finds out. And then she's going to blame us."

"So don't tell her—please." I pleaded knowing damn well they would tell Mom and she was going to kick my ass to Zacatecas and back.

"I'm parking. We'll meet you inside."

Damn. *Lola, why?* Again I walked around the huge hall and right when I was about to start crying, I saw one of the apron-wearing *tías* in a corner behind one of the tables looking like she was trying to lift someone off the floor. And then I saw Lola's drunk face tilted back against the wall where she was sitting up. Thank you, Elvis! I ran over and saw that Lola was totally messed up.

"Estaba abajo de la mesa—dormida." The little *tía* let me know.

"Lola, why were you asleep under the table, man?!"

By this time, Lola looked like Linda Blair from *The Exorcist*—she was all tore up. Her hair was messy and partially wet, her

mascara was smeared, and she had a scratch above her left eyebrow. I was expecting split pea soup to come hurling out of her
mouth any second, but luckily, no possessed projectile vomiting.

"Tired . . . tired . . . couldn't find you . . . want to go back to
Mimi's." Then Lola slumped forward. As I tried getting her off
the floor, Eleanor and Marty arrived and helped me heave the dead
weight up on her feet. "Ouch, my hair! You're pulling my hair!
Ayyyy—quit it!" Lola whined.

All three of us dragged her drunk ass out of that place and just
as we were stepping outside, Tony walked by and stared at my idi-
otic friend. I tried to smile but I felt real embarrassed. "She enjoyed herself and so did I. See you on Monday. Congratulations,"
I offered.

"Man, if I was you, I wouldn't go to school on Monday—or
ever."

Marty always had the smart-ass things to say.

We shoved la borracha in the backseat of Eleanor's Falcon and
split. By the time we got to the house, Lola was completely passed
out. I went inside first to make sure my parents were asleep. They
usually went to bed early. "The coast is clear," I whispered and
propped opened the door. "Let's hurry up and do this." I plunged
into the backseat and scooped up Lola as Marty and Eleanor
grabbed her arms and legs.

Before we lifted her body, Eleanor declared, "As soon as she's
through the door, we're taking off to spend the night at Norma's."
I instructed them to pull on the count of three and with that, we
removed Lola from the car. We got her up on her feet and my sisters placed Lola's arms around their necks while I stood in front of
them guiding all three into the house. "Lola, walk, *mensa*! Come
on, you have to help us." She actually was conscious and moving
her legs as her head rolled around like a rag doll. "Cold . . . I'm

cold . . . my *chi-chis* are cold!" she cried out, so I quickly covered her mouth with my hand. "Lola, shut up and walk! Come on, we're almost there."

"Hold her lips together," Marty said with a grunt, delivering the bag of cement to the doorstep.

My sisters walked her into the house, they passed Lola over to me by placing her left arm over my neck, made sure I had a good grip on her body, and then tiptoed out. As Eleanor was quietly closing the door, she smiled and whispered, "I'm glad I'm not you," gave a little wave and disappeared into the night. "I covered for you when you came home drunk," I pouted to a closed door. Yeah, I told her!

I stood there for a few seconds psyching myself up so I could get that booze smelling blob into my room. I took a deep breath and moved her forward, willing her to walk, and she obeyed. She was moaning and bumping into walls but still, we were getting her body to move in the right direction.

"Sshh," I whispered. "Help me, Lola, go into my room."

"*Ay,* my hair!" she whimpered.

"Lola," I tried covering her mouth but that was difficult. "Shut up and get your ass into my room," I shot into her ear.

"My hair hurts . . . *ayyy* . . . stop it . . . Am I purty?" I wanted to punch her lights out so she would shut up. We finally reached my room that was at the opposite end of the hall from my parents'—thank God! Then I threw her on my bed, took off her boots, and placed a blanket on her.

"Do not wake up and do not puke on my bed," I commanded. But she fell asleep the minute her head hit the pillow. I was so exhausted. All I wanted to do was sleep for two days straight.

The sunlight hit my closed eyes like an oar across my forehead. I quickly covered my face with the blankets. "Okay, you girls. It's seven." My mom was on a rampage as she opened all the blinds. "If you are old enough to get drunk, then you can get up right now and go to nine o'clock Mass. Don't think you are going to sleep all day." Oh, the motherly love.

It took me a while to open my eyes because they felt like sandpaper and my mouth felt like the desert. It was the worst hangover I had ever had. I craved a bucket of Coke with crushed ice and lemon. I stuck my head out. "Mom." I sounded like the Cookie Monster. I cleared my throat. "I'm sorry, I swear, but we can't go to any Mass. I feel really sick and look at Lola; she's still fast asleep."

"I knew I shouldn't have let you go. Your sisters told me they left you for forty minutes to go get some burgers and when they came back you guys were all *pedo*!"

"But, Mom, Eleanor and Marty didn't—"

"*¡Callate!* How dare Mrs. Esparza let young children drink. I am going to call her in a little while and let her know how I feel!"

"Mom, please don't embarrass me. I'll clean every room in this house for a week—just don't call her."

"*¡Como, qué* week! You're cleaning this house for a month—maybe even a year!"

"Whatever you want I'll do, but please don't call her."

She stared at me for a while. I hated when she did that. "I don't understand why you have to be so disrespectful to me and your father." Ah man, I also hated when she gave me the disrespectful line.

"Okay, you don't have to go to the Mass at nine but you're going to the 12:15 and you're going to confession—*y ¡ya!* I already called Lola's mother and she said I could do whatever I wanted to you both—whatever!" And she slammed the door. A second later she popped her head in again. "Oh, and your daddy brought *pan dulce* for you kids. So you better eat it!" And she slammed the door again. There was fire in her eyes for sure.

* * *

I sat there poking Lola in her cheek to be annoying and to see if she would wake up but she didn't. She slapped her face and grunted. I wondered what my full punishment would be. I wondered if all this would have happened if I had worn that itchy dress. Just then the door swung open.

"And you're not going to your Junior Prom!!" Then that mad woman slammed the door shut once more. Geez, Mom. That's *two whole years* from now.

Fine, I thought, that's too girly anyways.

The Dreamers

Quince Never Was

BY Felicia
Luna Lemus

Fog machines filled the darkened ballroom with so much purple sickly-sweet haze that none of my family saw my grand entrance. I arrived at my quinceañera reception under a veil of darkness as my court formed two lines and bridged their hands for me to walk under. I emerged onto the dance floor, alone and unescorted, with perfect timing as the DJ started spinning the song I'd chosen for my first dance—Chopin's "Funeral

March." Did my father waltz with me in the quinceañera tradition? No. Estranged from him and not the sort to want to dance
with fathers anyhow, I danced alone, my arms floating in the air as
I spun slowly and happily, my neck lolling from side to side.

My family truly didn't even realize I was there until I traipsed
to the edges of the parquet dance floor and accidentally crashed
into everyone seated at the table of honor. As I struggled to regain
my balance, my fingertips grazed my grandmother's shoulder.
Frightened, she yelped and threw her hands up to cover her neck,
five thick fingers wrapped around either side to protect her jugular
veins. Right then I grinned extra big just for her—she'd always
said I was too gloomy, that I'd be so pretty if only I'd smile—and
the bone-white pointed tips of my custom elongated and sharp-
ened eyeteeth picked up the table's candelabra light and shone
brilliantly.

My teeth. Normal girls wanted quinceañera gifts of jewelry,
fancy stereos, or maybe even a car if they were really rich. Not
me. I'd begged for cuspids Dracula himself would envy. Months
and months of pre-quinceañera tantrums and adolescent coyness
later, my newly capped choppers shone as the perfect accessory to
my Goth quinceañera ensemble: a black silk gown worthy of
Queen Victoria's mourning attire—full bustle, draping black lace
cuffs, and high stiff black collar—accessorized with super-pointy-
toed black suede calf-high witch boots and cobweb lace gloves.

If I'd gone to a salon to get primped before my ceremony, I
would have taken along my Siouxsie and the Banshees *Through the
Looking Glass* poster and I would have said, "I want to look like
this, but maybe more spooky, please." I didn't go to any salon; my
big day grooming was all home-done, but it was perfect. Kohl-
lined eyes peeked out from under the most inky black, most fabulously ratted nest of hair ever; a smear of dark red lipstick covered

my lips; and I had on so much white powder that I resembled a Día de los Muertos altar sugar *calavera*.

The engraved invitations I'd distributed to my closest friends read:

YOU ARE CORDIALLY INVITED TO MY QUINCEAÑERA. CHECK YOUR SOUL AT THE DOOR.

At the end of Mass earlier that afternoon, I'd bowed before our church's Virgen de Guadalupe statue and respectfully left Our Lady a bouquet of fifteen flowers—deep orange funereal marigolds and lush carnations dyed black, all bundled together with a steel gray velvet ribbon. La Virgencita had winked her waxen eyes at me and purred, "Eat a slice of cake for me."

Who was I to disregard such commands? At the reception, dry ice smoldering in the crystal punch bowl on the black-velvet-draped buffet table in front of me, I carved out two slices of the fifteen-layer black-cherry-iced red velvet cake for myself as all my guests stood around me and cheered and tried to pretend, for my family's sake, that this was the most lovely quinceañera they'd ever attended. A bite of cake coating my mouth, I looked over at my grandmother and smiled again. To be completely honest, I sort of hissed a little. She saw the bloodlike smear of frosting on my teeth and I guess maybe for a second she thought I'd actually finally snapped and bitten someone on the neck, because before I knew what was happening, she'd fainted outright. Guests rushed to her side and fanned her with their party favor faux-raven-feather hand fans, and I started in on my second slice of cake.

"To us, Virgencita!" I toasted with a swig of bloodred punch.

Being undead had never tasted so sweet.

* * *

When I was a kid, my most prized possession was a delicate little porcelain figurine of a unicorn with a gilt horn and hooves and four cornflower blue flowers painted on its neck. The unicorn was tiny enough for me to hold in my hand, which I didn't do very often because I was horrified I might break it. Instead, I kept the unicorn in a small musical jewelry box my grandmother gave me. The unicorn lived in the jewelry box's main compartment, on a cloud of cotton puffs I'd pilfered from the grown-up stuff in the bathroom. Unlike most kids, I wasn't allowed a pet—the unicorn was my confidant and comfort instead.

When one of my best friends in kindergarten, Jennifer, asked me, "What's your favorite animal?" I answered, "Unicorns," without a moment's hesitation. Jennifer replied that unicorns weren't real, so they didn't qualify for favorite-animal status. "You have to choose something else. Come on, what's your favorite animal?" Jennifer had silky blond hair and lots of freckles—features that were the definition of pretty as declared by our peers and all of mainstream culture at large—and a name that slipped off our teacher's tongue with great ease. Jennifer could never have understood how important unicorns—those magical, kindly, exquisite creatures of flight—were to me.

Real unicorns, of course, always remained just beyond my reach. My childhood hopes to lay claim to a living, breathing unicorn all my own filled my chest with such frenetic hope that a pediatric X-ray would have surely revealed electrical surges sparkling like trillions of chewed Pop Rocks where my lungs and heart should have been.

Back to topic: My quinceañera might as well have been a real unicorn.

Not only was my family *not* Catholic, they were also staunchly

opposed to all things Catholic—including quinceañeras. Conse-
quently, I never actually had a quinceañera. But *if* I'd had a
quinceañera, it would have been safe to assume the following: I
would have campaigned for a Goth theme, and my grandmother
would have passed out at some point during the festivities.

Politely stated, my grandmother wasn't particularly fond of
my personal style as a teenager. Looking back on it, I can't blame
her all that much. Our family did have a public image to maintain,
after all. Well, at least in a small-town sort of way. Since the early
1900's, my family had lived on the same block of our little Southern Californian barrio—Cypress Street in the City of Orange, to
be exact—and they owned the barrio's only grocery and sundries
store, as well as several of the houses in our neighborhood. My
family's influence and position may not have been much on a
global scale, but it was considerable in our sadly rundown barrio.

Back in the 1980s when I was growing up on Cypress Street,
trees, let alone acres of tall cypresses, were an uncommon sight,
but the OVC (Orange "V"arrio Cypress) gangster boys' graffiti
and gunfire wasn't. All things considered, Cypress Street was a
very tight-knit community. As a kid it seemed that everyone on
our block—whether actually related or not—was my aunt, uncle,
or cousin. Take, for instance, my "cousin" Teresa.

When I was little, Teresa lived with her grandmother, who
happened to be my grandmother's best friend. At that point in my
life I, like Teresa, also lived with my grandmother. My mom,
working long hours at a nearby university hospital to complete
her medical residency, typically came home well after I'd gone to
sleep and left before I woke. And so my grandmother, with her
distinctly gruff love and overly strict supervision, was the adult
who got me ready for school, watched over me in the afternoons,
fed me dinner, and told me to brush my teeth before bed—she

kept tight watch of me from sunrise to sundown. Therefore, when she'd visit her best friend, I'd come with her. And if Teresa was home when we'd go visiting, we girls were told to go play in Teresa's room.

Teresa was three years older than me, pretty and poised, and I worshiped the ground she walked on. I was probably somewhat equivalent to an unpaid babysitting nuisance for her, but she was always kind to me. During one visit in particular—I was a tangle-haired scruffy five-year-old and Teresa was a tall and long-limbed graceful eight-year-old—my "cousin" must have been feeling particularly charitable toward me. In fact, she actually seemed to sincerely enjoy hanging out with me that day. As we flipped through our sticker albums, trading scratch-'n'-sniffs, Teresa sang along to her new Michael Jackson record, *Off the Wall*.

> *You can shout out all you want to*
> *'Cause there ain't no sin in folks all getting loud*

When I reminded her that our grandmothers would actually, contrary to the song's promise, be very upset if we shouted like the song told us to do, she just laughed and kept singing, adding a signature Jackson dance move to the impromptu performance with perfect ease.

Clearly I was a total dweeb of a baby, but I knew enough to realize how cool Teresa was. In fact, just being near her made me giddy and dizzy. I wanted to impress her so badly that once she finished singing, I offered her entire sheets of my very best stickers—hologram unicorns, cotton candy scratch-'n'-sniffs, fuzzy kittens—in exchange for some of her most plain and useless stock, those freebie stickers Scholastic Books used to give to kids who bought books at school. But even though I was blindly willing to swap for sub-par stickers, Teresa was a nice kid, a good girl, and she politely refused to take advantage of me in the economy of decorative adhesives. I'll never know her reasons for doing what she did next—maybe my awestruck admiration inspired some sort of reciprocal generosity in her—but I can tell you I was absolutely delighted when she reached under her bed to retrieve my special gift: a color photograph portrait of her First Communion.

I'm sure this doesn't seem like such a big deal. I mean, families order those cheesy portrait studio perforated sheets of photographs by the hundreds, same as annual school photos, and push a print on everyone they know to brag about what a cute and polite or at least passably acceptable and well-combed kid they have, right? Probably. But considering my family was so damned anti-Catholic, for me the photograph portrait of Teresa's First Communion was a wallet-sized bit of contraband. Truth be told, beyond gut certainty that my family wouldn't approve of a Communion for me, I didn't really know what a Communion was. Teresa had probably mentioned something or other about her Communion during previous visits, but even with visual documentation of the event, the concept remained incredibly abstract to me. Still, the photograph, which Teresa signed the back of with looping little-girl cursive and a heart for the "a" of her name, was powerfully intriguing.

Add to the mystique, Teresa looked like royalty in the photo. There she was, posed daintily on her knees, hands raised below her chin in prayer, her big brown eyes dreamily gazing heavenward. Her expression was angelic, serene, and ethereal, but what struck me most about the portrait was her outfit: a pristine and almost blindingly white knee-length dress of spotless shiny satin with huge puffed cap sleeves trimmed in lace, a lace veil pinned and draping over intricately French-braided hair, white satin gloves, white lace-trimmed ankle socks, and white patent-leather

Mary Janes—the sorts of things I would have absolutely torn and sullied in under half an hour of wear, but that Teresa, of course, could be trusted to wear indefinitely.

"You look really fancy, like a bride," I said.

"Brides don't pray," she said.

I never ceased to be awed by her worldly knowledge.

Grubby little copycat that I was, I was overwhelmed with the sudden desire for a special day like Teresa's captured in color photograph portraits all soft focus and blurry around the edges like little clouds. As I stared at the photo, Teresa told me all about her First Communion, how it had changed her, how she was blessed by it. I listened attentively and, although I was certain I didn't want to commit myself to God, pay homage to la Virgen, or promise to maintain my purity, whatever the hell that meant, I knew I wanted a First Communion . . . along with the all the accoutrements, of course. After I hid the portrait in my sticker album so my grandmother wouldn't find it, Teresa said, "If you think my communion dress was fancy, just wait until my quinceañera."

"Your what?"

"My Sweet Fifteen. You can be in my court."

She might as well have been speaking Martian.

"My friend's older sister just had her quinces. It was a Cinderella theme."

Teresa went to her dresser and picked up a miniature clear "glass" slipper. She brought the cheap plastic shoe over and showed me the small lacy bundle tucked in the hollow of the slipper. With a delicate tug on said bundle's pink satin ribbon, Teresa revealed a small handful of what appeared to be a pirate treasure of silver inside.

"They're almonds. You can have one," she said.

I'd never seen such a thing. *Silver* almonds? Incredible. Deeply honored, I plucked one almond from the little mountain of shine, held it in the palm of my hand and I'm humble enough to admit petted it gently as Teresa spent the next hour telling me about her dream quinceañera.

Her face flushed and her voice nearly trilled as she explained her quinceañera would be *Flash Gordon*–themed. She whispered as she offered the top-secret information that her dress would be exactly like Princess Aura's was in the movie—a sexy gold lamé bikini top encrusted with diamonds and a long matching skirt with a slit so high one leg would be entirely exposed. And in place of a tiara like most girls wore at their quinceañera, Teresa planned to have an intricate, tall crown with tons of jewels and upward-reaching spikes.

Wow. I could totally picture it; if I squinted a little and imagined Teresa with a more grown-up body, she *was* Princess Aura. Forget a First Communion, I couldn't wait to have a quinceañera exactly like Teresa's!

Of course, although we didn't realize it then, there was a tragic cultural contradiction to Teresa's dream quinceañera. Whereas for the adults of our barrio quinceañeras were opportunities for United States–born girls to honor the cultural traditions that we—with our American schooling, snacks, and television, not to mention our ever-more-*pocha* tongues—constantly ran the risk of forgetting. In the eyes of the girls who had quinceañeras, the event was first and foremost an extravagant party thrown in their honor. Teresa was no exception. The quinceañera she dreamed of was the epitome of the materialistic, egocentric, and irreverent "American" culture our elders constantly reprimanded us for embracing.

Later that evening, once my grandmother and I had returned home, I tucked the silver almond Teresa had given me in my musical jewelry box, right next to my little porcelain unicorn figurine, and I dreamed of the day I'd get to be a princess too.

***

As chance would have it, within a year I was crowned a princess, albeit of an entirely different sort. My mother, long divorced from my father, remarried, and I suddenly found myself living several towns away in a new house complete with a pool, Jacuzzi, and two upwardly mobile physician parents. My stepfather drove a Porsche. I wore Esprit and Guess. My mother took me to get my hair cut at Vidal Sassoon in Beverly Hills. Each year I received an annual pass, laminated and bearing my photograph on it like a little driver's license, for Disneyland.

Although glossy surface appearances are often deceptive, from an outside perspective my life seemed filled with all things new, joyous, and magical. Everything I'd left behind at my old barrio home—my grandmother's watchful care, playtime with Teresa, my little jewelry box with the unicorn figurine and the secret silver almond stash—became part of my past.

The irony was that, by the time I was twelve—the drastic shift in my life firmly established and my borrowed dreams of a *Flash Gordon* quinceañera a distant memory—I ended up lusting after a religious rite of passage all over again . . . but this time it was a bat mitzvah. Had my family found religion? No, but in a way I had. Like most any other American twelve-year-old, peer conformity had become my God. And by the time I was in junior high, my friends—all of whom were wealthy and none of whom would have lasted a single day in my old neighborhood—included a handful of kids who began to celebrate bat and bar mitzvahs.

The ceremonies I attended were consistently held at reformed temples and the receptions were always thrown at posh hotels. I was confident that, if given half a chance, I could have gone to Saturday school for a year and learned a Torah portion same as my otherwise unobservant friends had. But honestly, more pressing on my mind was the fact that I, like my closest friends, wanted to be the kid who had the bat mitzvah party to top all others.

No joke, each reception I went to was identical in that all the earnest reverence and adult control of the temple ceremony was totally overthrown as we—gangly, brace-faced, spoiled, and raging-hormone-horny—ran wild in our pastel Jessica Mc-Clintock dresses and rented tuxes. To a soundtrack of Tone Loc's "Wild Thing" and Guns N' Roses' "Every Rose Has Its Thorn" we snuck gulps of booze, huddled in ornate hotel lobby bathrooms, choked on menthol cigarettes we'd bought from hallway vending machines, somehow managed to rent hotel rooms without our parents knowing, made out like crazy, and terrorized innocent hotel guests on the elevators—in short, we *totally ruled*.

Can you blame me for telling my mother I wanted a bat mitzvah?

For obvious reasons—I wasn't Jewish—my catered and très memorable bat mitzvah never happened. As far as I was concerned it worked out just fine anyway—my social calendar was booked with other people's celebrations and I never had to write thank-you notes for gifts or embarrass myself in front of everyone as I stumbled through phonetically pronounced Hebrew in temple like they did. For an entire year I, with my friends, rocked on. I would have partied even harder if I'd realized how rattling the transitions of the coming year would be.

At fourteen, I found myself far from all my junior high friends, back in the barrio, and sleeping on my grandmother's couch again. My mother, in the midst of what had become her second marriage's very ugly divorce, was gone at work most the time and left my grandmother in charge of me like the old days. I hadn't lived at my grandmother's house since I was six, but there I was, frustrated anew by her strict rules and "sensible" displays of affection. Gone were the days of my mom handing me a few fifties and dropping me off at South Coast Plaza Mall with my friends, where we'd shop and then sneak off across the street to TGIF's to flirt with the waiters who'd bring us pitchers of piña coladas served with far more rum than mixer. No more were the unsupervised closed-bedroom-door visits with a particularly cute skater boy who'd sneak his hands to places that made even a skater boy blush.

Instead, living with my grandma situated me back in the protected confines I'd inhabited as a kindergartener, with me pleading to please be allowed to ride my bicycle to friends' houses like all other kids seemed to do as naturally as breathing. Given that she didn't allow me nearly enough rope to hang myself, my grandmother's most constant and concrete complaint about my teenage presence in her house was that I tied up the phone line too much. This, of course, was only the surface issue. What actually upset her was that, due to recent developments, all my new high school friends, same as me, draped themselves in black clothing and made a hobby of dragging themselves around melodramatically like the hormonally imbalanced depressives they were.

"What, you want to look like zombies?" my grandmother asked.

Not zombies exactly . . . more like Bela Lugosi or Irma Vep.

If I had been the normal sort of kid who bought a yearbook my freshman year of high school, I'd be able to open said photojournalistic document right now and tell you about my vampire peers' appearance with biting specificity. The year was 1989 and if my memory serves me well, there was lots of aerosol hairspray . . . and dark lipstick, plastic tortoise-shell compacts of drugstore face powder in the palest shade available, cat-eyed eyeliner applied with ornate flairs such as little cobwebs reaching outward toward our widow's peaks, black clothes from head to toe, and black suede boots with long pointed toes and at least five buckles reaching up to torn fishnet-encased calves. But that was only what we—the drama-geek and often nerdy Goths, boys and girls alike—wore. As for the normies—the jocks, preppies, cheerleaders, and miscellanea of plain Jane and Joes—I have nary a vague recollection of how they adorned themselves. But whatever the normal kids looked like back then, I wasn't one of them. Hence, my grandmother said:

"No, I don't care if your *compañera's* older sister had her license *twenty* years, you aren't driving to the movies with those ghouls."

She was only doing what she thought was best for me, but it killed me that she thought I'd turned out so badly. There remained a part of me that wanted nothing more than to please her, to show her I was a good kid. And when I thought of good kids she gave her stamp of approval to, memories of my cousin Teresa as a daydreaming sweet young girl flashed in my brain. It must have been this combination of triggers that pushed my tongue to tell my grandmother I really wished I could have a quinceañera.

"Ha!" she said, taking a good look at me with a sweeping stare.

"Fly away, little vampire, fly away," my mother teased the couple times I pestered her about the idea.

But, unlike with my grandmother, I could see something in my mother's eyes that seemed to register the sincerity of my request. I wasn't sure how my mom was going to pull it off, but I hoped she'd somehow manage to throw me a surprise quinceañera.

The most traditional families in our barrio hired mariachis to serenade their daughter from the front yard the night before her quinceañera. I worried that if my mother actually went so old-school as to hire mariachis for my quinceañera, the mariachis would stop mid–"*Las Mañanitas*," pick up their instruments, and take off running when they caught sight of me. There I'd be—opening the living-room curtains to watch them sing—looking ever more frightful than Lily Munster before her morning coffee. My perversely hopeful concern never had the opportunity to be tested, because as my fifteenth birthday approached, rather than bringing mariachis to our front yard, my mother arranged for Robert Smith—lead singer of The Cure and most princely Goth heartthrob—to serenade me instead. Well, sort of.

Okay, allow me to back up for a second here. Let's talk about The Cure. To this day, hearing one of their songs sets free in my ribcage all the crashing intensity, confusion, and desire that nearly overwhelmed me on a minute-to-minute basis each and every day of my teen years. The Cure—purveyors of the saddest, most candied, and haunting melodies. Sure, they sort of sold out and went mainstream. Yes, hardcore Goths said they had always been way too pop to really count as Goth. But, whatever, The Cure was the band we most commonly included in the mix tapes we made for each other, their music was what played on our boom boxes as we cried and licked our frequently wounded hearts; their lyrics were the life vests we clung to as we nearly drowned amid our stormiest secret longings and vulnerabilities. Basically, The Cure might as well have been bottled in tonic form for the way we consumed it to soothe our teenage aches and ailments.

Hence the frenzy that ensued among my friends when KROQ—the Los Angeles radio station that, in the late 1980's, played songs we "alternative" kids dug most (KISS FM was for Top Ten pop losers, K-Earth was for dorks who liked oldies, and KNAC was for metal heads)—announced that The Cure would be touring with the release of their latest album, *Disintegration*. Adding to our obsessed delirium were the rumors that the *Disintegration* tour could be the final Cure tour . . . ever! The mere mention of The Cure's possible demise was enough to unleash torrential rivers of mascara-blackened tears down our collective powdered cheeks.

"I'll *die* if they break up!" we took turns sobbing in the hallways at school, slumped against our lockers and drawing even more than the usual quota of disgusted looks from our classmates as they walked by.

"I can totally feel my heart breaking. Seriously, feel!"

(Anything was an excuse to be groped by one another.)

To look at us, you'd think we'd welcome the possibility of death caused by unrequited desire, but no. Whereas hyperbolically performed melancholy was our trade in stock, for once our desperation and misery was real. Deep in our knotted guts and anxious lungs and bloodshot eyes, we ached. Mind you, wars were being fought. People were starving to death. AIDS was wiping out entire segments of the global population. But still, for us—the myopic, self-involved, and lucky inhabitants of a relatively safe and comfortable existence—the thought of missing what might be the last Cure tour felt like it might just extinguish our very souls.

"Mom, please, I have to go," I whined over and over and over again when my mother came home from work late that night.

"When is the concert?"

"Mom, pleeeeeeeeeeeeeeeeease."

"How much are the tickets?"

"Alice's parents are letting her go."

Ah, the charmed failings of pubescent/adult communication. Bringing up Alice right then probably only helped to convince my mother that Alice was exactly the sort of girl she didn't want me hanging out with. Unfortunately, the other things my mother knew about Alice included that she was allowed to ride motorcycles and wear dangerously short black leather miniskirts . . . to church—where her father was the pastor! I braced myself for a predictable whopper of a "if one of your friends jumped off a bridge . . ." parental lecture. Instead, my mom truly rattled me with her response.

"Fine, you can go," she said.

Whoa. What? Data does not compute . . .

"With a chaperone, Felicia. And I mean a parent."

The horror!

I didn't even bother telling my friends I could go to the concert. I would have been mortified to explain that my mother was making child-rearing decisions even my super-old-fashioned grandmother would agree with. So as my friends huddled around at lunch talking about how absolutely dreamy it was going to be to see The Cure, I kept my mouth shut. Two days into this conundrum, my mother surprised me with two concert tickets for field seating—the closest you could get to the stage. One ticket was for her and one for me, she explained.

"Tell your friends I'll drive everyone or they can meet us there."

Okay, so my mother was overprotective, but from my now-adult perspective, I see very clearly how thoughtful and generous this was of her. I mean, not only was she willing to take time out of her crazy long workday to stand in line at the mind-numbing hell of TicketMaster, she'd also secured an awesome ticket for me.

And now a pouting word from our fourteen-going-on-fifteen sponsor:

I did *not* want to go see The Cure with my *mother*.

"Never mind, Mom."

"Quit being such a brat; it'll be fun. Besides, the tickets are nonrefundable."

And, as it turned out, they'd been very expensive—as were the remaining tickets. The show sold out before any of my friends got their money together to purchase tickets of their own.

* * *

After I'd dressed and primped myself to perfection for the concert, I assembled an extra outfit of my clothes and shoes, and lay them out in the living room next to my mother's things. I was so not subtle, but what teenager is? Narcissistic creature that I was, I hoped my mother would at least *try* to blend in a little at the concert. The thought of her going in one of her business dresses with flesh-toned nylons and sensible shoes was enough to send shivers up my spine. Miracle of miracles, she took pity on me and played along without a hitch. In fact, she seemed to sort of revel in dressing the part, like she was preparing for some sort of titillating costume ball. Once she was dressed in her assigned ensemble of skinny black jeans, black button-down flowing shirt, and witch boots, I gathered all the necessary paints from my makeup case and called her into the bathroom. She unpinned her top-knotted long black hair and unbraided it so it draped to her waist in a natural veil. White powder over her already pale skin, her huge brown eyes heavily lined with kohl, burgundy lipstick applied with silent-screen-star cupid points—my mother looked enviably exquisite.

All through the night, I simultaneously loved and hated being at the concert with my mother. People kept stopping us to comment on how spectacular she was. You should have seen some of the intricate outfits and makeup jobs there, but still my mother made a memorable impression. Under all my white powder I was green with jealousy. I was just a typically insecure teen self-consciously aware of my every potentially dorky move. She was a self-assured woman of thirty-eight who looked like she barely even had to try to be cool. Okay, so I was impressed. My mom was cool. But, jeez, did she have to be *so* cool?

Oh yeah—the concert. It was awesome, I think. Truthfully, I barely recall what happened up on stage, but I do remember a group of fey pretty boys in the audience near us singing along and dancing admiringly close to my mom during the encore performance of "Why Can't I Be You?"

> *Everything you do is simply kissable*
> *Why can't I be you?*

When the show was over I made the huge mistake of mentioning to my mom that I wished we had backstage passes for the KROQ after-party.

"You need a pass?"

"Duh, Mom."

Pass, schmass, apparently my mother didn't need no stinkin' pass. She dragged a mortified me—*You can't just go back there without a pass, Mom!*—to the backstage gate.

"Hello," she said, smiling confidently at the security guard. "I hate to impose, I'm sure you're very busy, but I wonder if you might be able to help. You see, my friend has our passes, but I think she's already inside. My daughter can wait here for me. If you'd be so kind as to bend the rules, I'll find my friend and bring you our passes. . . ."

The fierce-looking security dude with his huge body-builder muscles actually smiled politely, opened the gate, and waved her in. He must have reasonably figured that no person as friendly, articulate, and rational as my mother seemed would leave her kid as ransom if she wasn't actually planning to come right back. And that's how it came to be that I waited backstage at Dodger Stadium, the big burly security guard watching me, for over an hour. No shit. My mother either forgot me entirely or was having such a great time that she couldn't be bothered. Wasn't this the same woman who wouldn't let me go to the show without a chaperone? Regardless, the security guard realized he'd been had and refused to let me in so I could find her. And, seeing as this was in the days before cell phones, there was no way for me to get in touch with my mother. Hoping the whole thing was a huge misunderstanding and that maybe she was trying to find me, I found our car in the vast parking lot and waited for her.

A cluster of rare Southern California thunderclouds gathered overhead as I sat on the hood of our gray Jeep Cherokee. Chin in hands, moping, I got thoroughly drenched as it rained. What a drama queen, right? Yeah, well, whatever, I *was* fourteen. People scurried by on the way to their own cars and worriedly asked if I was okay.

"Yeah, thanks. I'm fine. Totally fine."

I sat until my bony butt couldn't stand it any longer and then I walked back to the security guard to plead for mercy. As I stood still outside the gate, two white vans with tinted windows pulled out from the fenced-off backstage area and drove away slowly as clusters of frantic fans ran after them. The band. There they went. Bye-bye. I was soggy, agitated, hungry, and it was way past my bedtime on a school night. I started to bawl. The guard looked at me like I was the most pitiful thing he'd ever seen and he finally let me in, more to get me away from him than for any other reason, I think. I wandered around backstage, awkwardly calling out my mother's first name until I eventually found her. She introduced me to her group of new friends, which included some of Los Angeles's coolest music and radio insiders, and then said:

"Too bad you missed Richard. What a flirt!" She giggled.

Richard? Who? And then, with my next quickened heartbeat, I realized she meant Robert Smith, the lead singer of The Cure, my dream love! Of course, she would introduce me to him, right? My heart palpitated and my hands turned slick with nervous sweat. But then I remembered—the entire band had been escorted away in the tinted-window vans as I stood outside the after-party gates waiting for my mother! I held back sobering, bitter tears.

"*Robert*, Mom. His name is Robert. Robert Smith," I said.

"Oops," she said and giggled.

My mother and her new friends gushed on and on about absolutely nothing of any importance, and I suddenly saw the world with new clarity and understanding. I realized that deep down the only true difference between adults and me was that they'd had more years of living. My mother never said the concert was my surprise quinceañera, and she may have never actually intended it to be that exactly, but, same as any girl at her quinceañera, I went through a transformation that night. I became a young woman. One thing was certain, the night with my mother at The Cure concert was exactly as any quinceañera of mine would have been—melodramatic, bizarre . . . and strangely wonderful.

Over a decade later, as I was going through my old things in storage in preparation for my move from Southern California to Manhattan, I found the little musical jewelry box I'd kept at my grandmother's house as a kid. When I opened the jewelry box's flowered blue lid, tinny music started up and the plastic ballerina inside pirouetted stiffly. Therein I found three jewels.

A yellowed ticket for The Cure, Dodger Stadium.

A dainty porcelain unicorn figurine.

And a single silver Jordan almond.

I tried to bite into the almond that still shone all mirrored high-gloss like a low-rider car's hubcap, but I barely scraped the surface. Silver dusting my teeth, I gave up. The music box broke in the move. I lost the concert ticket. And something stronger or hungrier than me eventually ate the almond. But I still have the unicorn figurine.

In fact, the unicorn sits on my desk as I write this, demanding attention from the periphery of my vision—exactly as fantasies of real unicorns did in my childhood and hopes for a Goth quinceañera did in my adolescence. It's true I never claimed as mine a living unicorn or an undead quinceañera, but both have contributed to defining the adult I grew into nonetheless. And as for the porcelain unicorn, if you look closely enough, you'll see that each of its legs broke off accidentally at one point or other and have been glued back on. Sometimes I think I should buy a small bottle of gold paint to touch up its battered golden hooves and horn. Most likely, I'll leave the gilt chipped as a reminder of everything the unicorn has been through.

A Blue Denim Quince

BY Berta
Platas

I grew up listening to tales of extravagant quinces, the middle-class Cuban version of debutante balls. In my mind, the parties from the old days were like the cotillions that I read about in the Georgette Heyer novels I checked out of the New York Public Library. No minuets or scandalous waltzes, though: My vision was filled with the infectious beat of the rumba, and the women were dressed like my mother was in the fifties-era hand-tinted photos that she brought with her into exile.

I imagined myself wearing one of the sweeping, big-skirted silk illusion ballgowns that were in fashion all across the world when she was a teenager, thanks to designers like Givenchy and Balmain, and dancing with a handsome young man in a tuxedo.

The summer I was thirteen, the dreamy wishes became hard-core planning at a family gathering on a hot Miami night. Maybe it was my yearning for that long-lost elegance that made me so irritable as I sat between two overstuffed aunts, my *tías* by marriage, as they discussed my scrawny cousin Mirta's upcoming quince.

I winced at their tacky ideas, knowing that the perfect quince was a time to remember forever. I shouldn't have been surprised at some of their weird ideas; their kind of decorating taste ran to clear plastic upholstery covers over crushed velvet.

It was only ten o'clock—early—and I was sweaty, having just finished a marathon set of merengues with my uncle. Midnight was hours away, and with it the end of our family party. The police always showed up, called by the stuffy neighbors across the street who, as usual, turned down my *abuelo*'s invitation to mingle with our boisterous clan. It was like a ritual. The Calling of The Cops.

I don't remember what the occasion was for that summer fiesta (which means there probably wasn't one, other than an excuse to get loud), but I do remember that I was two whole years away from being a quinceañera.

I sat at the dining-room table, which had been dragged onto the polished-concrete-tiled patio in back and was covered with plates of *bocaditos*—tiny, crustless sandwiches—as well as big bowls of ice mounded with boiled shrimp and surrounded by smaller bowls filled with my grandmother's fiery cocktail sauce.

I munched a *bocadito* and surreptitiously sniffed my armpits to make sure my deodorant hadn't failed. Reassured, I sat back to listen to my ample-hipped *tías* talk over my head as I watched my favorite dance partner, my *tío* Manolo, boogie with his mother.

There are seven Manuels and five Manuelas in the family, making for a lot of confusion. They were nicknamed Manolo, Manny, Manolita, and Manuelito—and that's just the men.

"*Mijita*, you wouldn't believe the dress Mirtica found in Fort Lauderdale." Tía Mercedes pronounced it Fo-do-dell-dell.

"For her quince?" That was Tía Linda. That got my attention. I'd been spinning dreams of my own fabulous quince since I was little, but had never been to one. First we'd lived in a Polish neighborhood in Pittsburgh, and no one in our Manhattan neighborhood knew what they were either. So when I heard folks talking about planning, or even better, attending, a fabulous quince, I listened closely, hoping to pick up pointers, pros and cons.

Like the folks who were so stingy that they invited only the girl's family. Imagine that. They alienated an entire neighborhood, all their family friends, and the poor girl probably lost all of her friends at school. My parents were understanding and practical. I knew they wouldn't fail me when it came to the guest list.

The *tías* were still talking about the dress from Fort Lauderdale.

"It cost *un ojo de la cara*," Tía Mercedes said, "but *Dios mío*, it's beautiful. The top is tight, and cut low, like this." She scribed an arc low over her pillowy chest. A round gold medal of the Virgen de la Caridad del Cobre the size of a demitasse saucer was wedged sideways in my *tía*'s sun-spotted cleavage. "And it has five tulle flounces. It's es-strapless, you know, and covered with little *rositas de seda*."

Strapless with silk rosebuds. Sometimes my brain felt mushy from all the instantaneous translation.

The dress sounded heavenly, and I suddenly hated Mirta,

though I hardly knew her. She was a distant cousin, always out with her older friends, all of whom seemed to have fancy cars, while my sister and I were trapped in Viejita Land with our parents. My cousin was fourteen and I felt as if I'd been thirteen for twenty years.

I couldn't wait to get my own learner's permit, and then my driver's license. I'd cut back on all the endless family visits. My Miami holidays would be spent at the beach, or cruising through the malls. Maybe I'd have my own quince here.

I imagined myself in Mirta's gown, could almost feel myself sweeping down a red-carpeted grand staircase, tulle skirts floating after me. Skinny, big-toothed Mirta would look ridiculous in that grand dress. And what would hold the strapless bodice up? Not her flat chest. Miracle bras hadn't been invented yet. That unkind thought totally ignored the fact that I was equally skinny, flat-chested, and had teeth like a piano keyboard. My imaginary self was curvy and elegant. Maybe I'd fill out in two years. Time was running out for poor Mirta.

That night, after more merengues, too much *puerco asado*, and a few stolen sips of Tío's beer (yuck) and Tía Chela's rum and Coke (yum), I dreamed of Mirta's gown, and awoke in planning mode.

My own quince loomed closer. Two years suddenly didn't seem like a lot of time. Soon I'd be the center of attention, surrounded by my own elegant court, and of course my dress would blow everyone away. I figured by then Mirta would be old, maybe even married.

Inspired by the description of Mirta's gown, I started to plan my own dream dress. I drew and discarded hundreds of designs. I'd been a fan of fashion since I was little. My mother was an expert seamstress, and she made clothes for herself and my sister and me. She'd gotten her hands on two huge McCall's and Buttericks catalogs, the ones that sit on slanted tables at fabric stores. I'd spend hours tracing the heads and arms of the fashion drawings, substituting my own dress designs for the ones in the book.

I knew that in the garment district we could buy any kind of cloth and ribbon that existed, and I envisioned that whatever I designed could be assembled right in our apartment.

My best design time was on Saturday afternoons, after house cleaning, when Mami would set up her creaky ironing board, with a cup of water on the wide end for sprinkling, and her steam iron plugged in. As she ironed piles of shirts, sheets, and even my jeans, I sat at her feet.

I had an old soup can filled with colored pencils, a bit of rolled-up newsprint for blending, five extra-sharp number two pencils for drawing, and a fat pad of newsprint sheets.

While Mom ironed, watching old black-and-white movies on TV, I would grab my pencil whenever I saw a dress I liked in one of the overblown historical films we both loved. They starred women like Merle Oberon, who seemed unbelievably romantic in big, poufy gowns. I imagined the dresses in wild colors, in virginal white, or pale, delicate pink. My colored pencils were quickly reduced to tiny stumps and my dad complained about my constant requests for new ones.

I argued that it was for art's sake, but kept my dress designs a secret from him, afraid he would think they were frivolous. As the oldest child I was my father's self-appointed "boy," and had learned to hunt, play baseball, and follow arguments about old-time Cuban politics. If he'd approved of women smoking, I'd probably have learned to smoke a cigar, too.

Not that he wanted me to be a tomboy, but I desperately wanted to please him, and without asking, I knew that fashion design would not be his choice of careers for me. I was destined for a greater, and more serious, career. Law, maybe. Or teaching.

Meanwhile, I ate up those classic films and blended them with the stories I'd heard of the lost Havana that the older folks painted as heaven on earth. A heaven filled with parties fueled by expensive drinks, twenty-piece orchestras, and fashionable, happy dancers. Outside, lines of limousines would wait to carry them home, their uniformed drivers speaking with their hands in broad gestures that were echoed by the men who smoked fancy cigars on the stone balustraded balconies above them.

I envisioned myself entering a crowded ballroom, the wide, diaphanous skirts of my gorgeous gown skimming the polished floor. Everyone would be silent when they saw me, and then a waltz would strike up and my father, dashing in his tuxedo, would lead me into the first dance while my mother looked on proudly, her eyes shiny with tears of joy.

After my father had twirled me around the floor, all my friends would join the dance, forming the traditional fifteen couples—one couple for each of my years on earth. Later on my escort and I would join them as the sixteenth couple. We'd be the most beautiful couple, of course. I'd have the best gown, the most handsome escort, and we'd be the most elegant dancers. A tiara would glitter in my dark curls.

Never mind that by then we were living in Charlotte, North Carolina, a city that in the seventies was divided like a bakery cookie, black on one side, white on the other. I don't know what that made me—the creamy filling? But there were no other Latinas anywhere, or else they were in serious stealth mode. When I ran into a girl whose last name was Diaz I got excited, only to find that she pronounced it Die-ahz. How can a family forget how to pronounce their own name? If she had a Latino in her past, that person's blood had long ago surrendered to the rest of her family's Irish DNA.

The lack of Latina girls my age did not deter me from my quince plans. I'd heard of parties where the *festejada* was such a troll her parents had to rent her *corte*, and their dates, too. Pitiful. That wouldn't happen to me. I had lots of friends, although all of them were Anglos. I'd have to compromise.

They had some compromising to do, too. The students at my junior high were mostly African American, which was new to me. I'd gone from the Polish neighborhood in Pittsburgh to the Hungarian Jewish/Dominican/Puerto Rican one in New York, and they had all been blends of immigrants. We'd had a lot in common. To me, Charlotte was like a foreign land. They probably thought the same of me. A Cuban. From New York City, no less.

Although my new friends saw fifteen as a time of huge change, it had less to do with coming-of-age parties in a ballroom and more to do with getting a driver's license. While my girlfriends studied the rules of the road, I wondered if ribbon roses would be too tacky, and should we serve beer or stick to wine and punch?

My girlfriends would have to learn to do the dances, and they'd need pointers on what to expect. There was still a risk of extreme culture shock as New South met Old Cuban.

We'd all had a little taste of it already, when two carloads of family arrived from Miami, the likes of which our little neighborhood had never seen. Long before the influx of Latinos from Mexico and Central America, we were the only ones any of these blue-collar Tarheels had ever seen.

The sight of my entire family arrayed in the front yard after dinner, seated in webbed aluminum chairs and holding six simultaneous, loud conversations was an eye opener to people who kept their family lives private and confined to their chain-link fenced backyards. We noticed that the traffic on our quiet street increased, as neighbors found excuses to drive by, as if we were a zoo exhibit.

But my friends Ellen and Sue were not amazed at the amount of pork, beans, rice, and sweets my family consumed. They were Southerners, after all, and their own families were large (in more ways than one).

But the strong Cuban coffee that brewed in the silvery stovetop *cafetera* night and day intrigued them. And they were shocked that even the children drank it, mixed with hot milk, lots of sugar, and a pinch of salt in a *café con leche*. Their eyes widened again at the sight of all the eighteen-karat-gold jewelry looped around my family's necks and hanging from our ears. They were especially intrigued by the holy medals that the men wore, dangling from thick chains. Some of those medallions were big, like miniature Virgin Mary hubcaps.

Ellen and Sue got more into my party planning after that visit. Although they still rolled their eyes when I talked about waltzing, they were totally into the dresses. I was thinking Old Cuban Tradition, and they were thinking *Gone With the Wind*, which was probably closer to the truth. Gone With the Communist Wind, maybe.

At least I didn't have to worry about my escort. The summer I was fourteen, a Cuban boy moved into my neighborhood. Andres was darkly handsome and very shy. His family had moved to Spain instead of coming to the United States, so he was very different from any Cuban I'd ever met.

His old-fashioned manners seemed alien to me, as if he'd landed in Charlotte by way of a time warp, but at the same time, he seemed a lot like the heroes of the movies I used to watch.

I fell madly in love with him, and better yet—I'd found my

escort for my quince. Who better? Now my daydreams featured Andres, who would take me from my father's arms after the first waltz, and spin me around the room, skirt swirling. Despite the passion of my first love, that dress was still just as important as my date. Maybe more so.

A dreamer, but a pragmatic one, I planned carefully. This waltz here, that teary moment with my father there. The flowers, the food, the music. Friends made it onto my list, or were angrily scratched off after a fight.

My mother shook her head and called me *loca* when she saw the sketches for my gowns. She'd always said that there would be no big quince party for me. For one thing, we were in the United States, where no one did stuff like that. For another, we were poor and couldn't afford it. That last one was the real reason. I ignored her.

Of course we could afford it. How much did a party like that cost? A couple hundred dollars? I began to do my homework. At the bakery I scoped out the biggest cake and asked, how many would it feed? How much did it cost? The answer was shocking.

And when Ellen's older sister got ready to go to her prom she showed off for us, whirling around in her sleek red halter dress. I loved it, and it was perfect for her prom. But not for my quince. Too sleek, too contemporary. Her date was wearing a fabulous tuxedo. Ellen was mortified when I asked him how much the rental had cost, but I didn't care. Andres would need to know.

To help my mother out, I left a list of all these expenses on the breakfast table. I figured I'd done all the homework. There were no palaces or mansions available for the venue, but the Hilton downtown had a great ballroom with a huge dance floor. I mentioned it in my note. The First Baptist of South Park also had a really decent hall, although the possibility of dancing there was iffy. But if I was ready to compromise, maybe they could, too.

I waited and waited for Mami's reaction. I figured she'd want me to whittle down the guest list, or choose a smaller hall, but I didn't expect total silence.

By dinner I couldn't wait any longer.

"What did you think of my guest list, Mami?" My strategy was to start off the conversation with the thing she was least likely to change.

She stopped chewing her mouthful of *fricasé de pollo* and glanced at Papi. Not a good sign. Papi's eyes never lifted from his plate, which meant nothing, since he was single-minded about his meals.

"I saw your little note, *mija*. But we've discussed this before. There's no money for a big quince. Don't worry, though. We'll do a little party for you."

A little party? Whatever else she said was lost in the roaring that filled my head. I imagined storming away, slamming my bedroom and screaming, but I knew that wouldn't get me the party I wanted. I swallowed my intense disappointment and decided that I needed a more subtle approach.

My parents were probably going to pay for my wedding someday. Why not spend that money today? I could come up with the cash for my wedding, although they'd still have more time to save up for it. I wouldn't get married for years, right?

And what if I never got married? What a waste of resources! They could pay for my quince now, and never have to worry about upsetting me later.

My fifteenth birthday approached, but I saw no sign of any big surprise party. Andres consoled me by teaching me to French kiss, and practicing did wonders for making me feel better about the lack of engraved invitations and the Hilton ballroom left empty, undecorated. Maybe it would be a family celebration. So what? I could show off my handsome boyfriend to all of my cousins. There was still time for the big dress and the Baptist hall.

When my birthday was only a couple of months away and still no plans were forthcoming, I prepared myself for a surprise party. Although maybe it wouldn't be in a rental hall and maybe there wouldn't be a big catered dinner. I scouted out locations in the backyard for the roast pig, and wondered if the band would fit on the screened-in porch. And where would everyone stay if they all showed up?

I kept my room clean and tidy, figuring that we could fit at least four of my cousins in my bed, in my sister's, and in sleeping bags on the floor. My aunts and uncles could get hotel rooms. I cut my party expectations back a few more notches.

With a pig roasting in a pit out back and lots of friends and festivities, it would be a real Cuban party. I'd just have to make do with making my appearance from the kitchen steps. Some entrance.

That spring, work and bill-paying took precedence over my quince. My father had taken a second job, and even our annual summer vacation to Miami was put on hold. My mother loved my final dress design and said that she could certainly make it, but she refused to buy the cloth. It would be too expensive, she said, and added, "Why would you want a dress you'll never wear, anyway?"

I went to my room, shut the door, Scotch-taped the drawing of my gown to the wall over my bed, then threw myself down, letting the sobs come out.

I'd have to accept the truth, that I'd never have the party of my dreams. I felt foolish and betrayed and very sorry for myself. I gathered all of my planning notebooks, the guest lists and rental hall brochures, and cried until my chest hurt. When I finally fell asleep, a tuxedoed Andres consoled me with kisses.

* * *

The morning of my fifteenth birthday, my father woke me up by turning on the bedroom lamp and putting a glass of orange juice in my hand, as he did every morning. This day I made room for him, and he sat on the edge of my mattress and kissed my cheek.

"I wish you could have had your big party, *nena*," he said in his deep, radio-announcer voice. "But we're not rich. And the rich people who have this kind of party, do you think they appreciate it? Better to have love and a little cake than a huge extravaganza. You'll see."

Right. Someday I'd understand. I'd heard that one before, and for certain I'd hear it again. I got dressed for school, where my friends greeted me with cards and threats of major spankings. No one there thought any less of me for not having a blowout bash. I felt a tiny bit better.

Back home, my mother gave me all of the envelopes that had arrived from my far-flung family. She always got the mail before I arrived home, so she was able to hide the cards and letters that had accumulated.

Everyone in the family had remembered my important day, and the envelopes were full of sonnets composed just for me, cards signed with lots of love and marked with kisses, and birthday checks written against banks located in all of the cities where the Cuban revolution had flung my ambitious kin.

My *gringa* friends were astonished. Their mothers had told them about old-fashioned Sweet Sixteen parties from back in the days of malt shops and big-fendered cars, but even then, no one got this much loot.

That night, we had cake and ice cream, and my four best friends joined my family, singing while I blew out fifteen candles. No one waltzed. My girlfriends and I wore jeans and tight-fitting long-sleeved tops with lots of long, beaded necklaces and big hoop earrings. We looked totally cool, and I smiled as if I wasn't heartbroken. I deserved an Oscar for that performance.

Afterward I sat alone on our concrete front steps, illuminated by the street lamp three doors down, and thought that without my quince party, it seemed like any other birthday. Only the quinceañera title made it different. I would be Berta, la quinceañera for a year. Just like a beauty queen.

Beauty queen? As if. I was still stuck in the same body, although now it could get a learner's permit. Legally, anyway. The thought of me behind the wheel frightened my overprotective parents.

The aftermath of my fifteenth birthday meant that my time with Andres was watched more closely now, as if all of a sudden I'd have the urge to do something that hadn't occurred to me before. Jeez. I had more sense than that.

Weird how the event that is supposed to signal the start of womanhood, with a virginal gown and weddinglike trappings, is followed by increased monitoring over my driving, dating, clothing styles, makeup, and friends.

I don't think that chastity belts will ever make a comeback, especially since they're probably god-awful unsanitary, but if any enterprising soul starts selling monitoring ankle bracelets for Latina girls, the market is practically unlimited.

To make it even more agonizing, when guys hit the same age they are allowed almost total freedom. Kiss your *mami* hello, plant one on your *abuelito*'s forehead as you stroll through the house. Ooh, you're such a good boy.

Girls can't just breeze through on the way to the bedroom, not if they're going through the kitchen where the women are fixing dinner. As if estrogen was glue, we stay with them, listening to family gossip, and the worst of it is, it eventually sucks you in. You start to enjoy it. You feel that this is where you belong, in the kitchen, the unacknowledged mistress of the universe.

But a quick walk through the house exposes the other side. The guys, slapping the table to emphasize jokes, sucking down iced beer, talking loudly, leaving their messes for the women to clean up. They think they're in charge, poor fools.

So the dilemma is: stand firm and be a modern woman and be doomed to listen to football talk and endless diatribes about what the latest president did wrong and how different things would be if only we were still in Cuba, or cave and go back to the kitchen, full of soothing talk, hysterical laughter, and pitchers of mojitos.

I'm telling you, it's a conspiracy.

Even in Cuba, a big quinceañera event was not the norm for my family. A raucous beach party would have been in order, not stately dances in formal clothes. But I learned later on that even a party at home could be a wild affair. This I discovered when I was invited to an extravagant Miami quince when I was twenty-one years old.

I won't say whose party it was, but I'll tell you about it. And I've changed the names to protect myself from a Coconut Grove beatdown.

Tania was the cousin of a cousin, which made her practically immediate family. Her sleek gown cost almost a thousand dollars, more than that if you added the shoes and the diamond bracelet that she received as a birthday present. She already had a BMW, so the usual car-or-party offer wasn't made. When I found out that this was the custom in Miami, I got momentarily depressed again thinking about my humble quince.

Tania's lavish uptown gala was segregated, adults on one side of the pool, kids on the other. She made her mandatory appearance to *ooh*s and *ahh*s, followed by the stiff, prearranged dance routine that was always the idea of a girlfriend who thought herself a good dancer. Thank God there weren't any opera-singing girlfriends. We'd have had to sit through endless arias.

Then there was her father, who'd already downed three vodka gimlets at the prospect of receiving bills for all this stuff, followed by another three at the idea that he had to get out there and dance with his little princess in front of everyone. He ended up tripping as he turned his Tania and they almost went into their Olympic-size swimming pool.

The pool had Greek goddess sculptures on either end, pouring water out of their urns. The urns poured and poured, and frankly, they made me want to pee, especially after all the mojitos I'd been polishing off. I have to admit I was disappointed that they hadn't taken a fall. A quick, wet finish to the formal part of the festivities would have been grand.

Instead, we got almost forty-five minutes of a stilted over-arranged dance number (I swear they did a sort of cumbia-minuet at one point), and then Alberto, Tania's dad, got all teary when he took off her little flats and held out stiletto-heeled jeweled slippers, genuine Manolos, symbolic of her entrance into womanhood. Those shoes cost as much as two of my car payments.

The women salivated and there was yet another chorus of *ooh*s and *ahh*s. The men glanced at their watches: The Marlins were playing that night.

The minute the shoes were on her feet, the party went wild.

Waltzes over, drum-heavy electronic music blared and everyone joined the dancing. The adults got steadily drunker and as their vigilance blurred with their speech, the kids joined in the drinking fun. The bartenders seemed not to care.

I danced until I was sweaty and unsteady and figured out that chugging mojitos as if they were minty lemonades was not the best way to cool off.

I joined the nondancing adults who were seated in a circle on the adult side of the house. Their conversation has been only sporadically interesting, and when they launched into their customary "the Cuba that was" lament, I crossed to the other side of the house to see what the teens were up to. Holy cow. Good thing the priest left early.

As I walked over, I spotted a jeweled slipper embedded in the stiff hedges, followed by moaning coming from behind the hibiscus, where no doubt something else was stiff, too.

The air was thick with humidity, the perfume of the flowers around us, and a little tang of salt. Or maybe it was spilled beer. The moaning grew louder, and I thought maybe not all the fun was consensual, and was ready to rescue whoever was crying out.

At the end of a gravel path, by an outdoor water faucet, I found the source of the moaning. Eddie Alvarado, the quinceañera's escort, was doubled over in his rented tuxedo, puking. It smelled like rum and sofrito. I blew out the nearby tiki torch, afraid that fumes from the rum-soaked gravel underfoot might ignite. Then I turned the hose on and let him rinse off his shoes and swish out his mouth with the tepid water. I wondered where his date was. I'd spotted one of her shoes, but the rest of the girl was nowhere to be seen.

I found the birthday girl in the home theater. Tania was pitching a force-ten fit, angry that no one was doing her bidding anymore. Her makeup was perfect despite the tears that ran down her cheeks. Kudos to the makeup guy.

The big, velvet-walled room with five rows of theater seats was empty except for two bored-looking teenage blondes, probably Tania's friends. Or former friends, from the smoking words that were pouring out of her ruby-red lips.

The curtains were pulled back from the wall-size movie screen behind her, which was showing a documentary-style film of Tania trying on gowns and tiaras and checking out different updos.

The lights were on so the picture was pretty murky, and whatever was being said was impossible to catch.

The real-life Tania's words were crystal clear. After weeks of rehearsal, a videotaped, choreographed dance, all the rituals between little girl and adoring daddy, she'd been dumped by her friends, who were all merrily dancing, playing video games, and screwing. All except for these two loyal followers, whose eyes kept straying to the door behind me.

It looked like a scene from a movie about the Old Testament, one of the biblical stories that was followed up by God's punishment and end badly, with a cataclysm of some sort. I backed out into the night before she saw me.

It was past midnight, and the poolside DJ had kicked the sound level up a few notches. Usher's latest tune bounced off the windows of surrounding houses.

I didn't care to meet any Miami cops, so I kissed the hostess good-bye and slipped into one of the cabs waiting out front. The next day there was nothing in the *Miami Herald* about the party, so I figured it had turned out okay.

Luckily, that particular quinceañera grew up, went to college, and got a job. No harm done. At least, none recorded here.

After that party, I considered myself cured of my missed-quince obsession, until a few months ago in Atlanta. My Anglo sister-in-law was in town and we went to a local bridal store, famous for its huge selection of party gowns. Our mission was to find a gown for her to wear to her son's spring wedding. We were looking for an elegant gown that didn't scream "mother of the groom."

The store was a sprawling warehouse jammed with gorgeous gowns—wedding dresses, evening gowns, prom dresses, and knockoffs of every Oscar-night red-carpet gown from the last ten years.

While my sister-in-law tried on dresses, I wandered around, agog. Slinky red. Virginal white. Poufy gowns in delectable shades of pink and black. Elegant and sophisticated, beaded, crystal-bedecked, flounced, crinolined—it was all here.

My quince dreams came flying back, like fairies out to prove that they were real.

I saw the perfect dress. It wasn't a prim and virginal gown. This one was black and corseted, with an immense skirt decorated with sprays of large, primary-colored metallic spangles trailing through the layers of black tulle like brilliant bubbles in a night sea. The sleeves were little wisps of transparent black, gauzy silk, with some black beads for sparkle.

I was totally in love. Forget the grown kids and the fact that my fifteenth birthday was decades behind me. I instantly reinvented my ideal quince to feature me in that dress. Between evaluations of mother-of-the-groom gowns in the dressing room area, I returned to look at that dress.

I knew I was totally old for that outfit, but I walked by, pretending I was bored. I ran my fingers through the paper tags on the dresses, letting the little strings tickle my fingers, the little cardboard price squares bounce against my skin. I turned it around. Five hundred dollars. I could afford it. They had it in my size.

The Asian saleswoman met my eyes and smiled kindly. I wasn't fooling her. How many women came in here for a matronly outfit and ended up wistfully examining a little frock made for a fifteen-year-old? When the saleswoman came out from behind the counter to offer advice and help my sister-in-law gather more gowns, I saw her look at me knowingly. Feeling pathetic, I confessed that I loved the dress.

She grinned. "That dress is very special. Very fun."

"What kind of girl buys this dress?" What I really wanted to ask was, "Can I try it on?" I yearned to wear it.

She laughed. "A self-assured one. A girl with a big sense of fun."

I had a big sense of fun. I like to think I could have pulled it off when I was fifteen, although maybe not in black. Not when I loved pink ribbon rosebuds.

I know I'll never own that foil-spangled black tulle dress. Where would I wear it? Maybe my daughter will have one like it one day, or her version of it.

She's an assertive little girl, and would look great in that black gown. But it will be her dream, what she wants, that counts. If she wants a quince, she'll have it, although I'm hoping she won't want an out-of-control debutante bash. I'd rather blow my budget on her dress . . . and the SWAT team that I'll hire to keep a close eye on the proceedings. I know kids do dumb and forbidden things and survive, but not while I'm in charge. And if she does anything crazy, I don't want to know about it until way after. Like, when I'm on my deathbed. Okay, so maybe I'll make her the car/party offer.

My quince was not the beautiful party that I'd dreamed of, but I have a lasting reminder of it. The cake is long gone, and we took only a few pictures, probably because I looked so glum. But the next day I went to the mall with my friends, armed with some of my family's expressions of love (in cash!).

It was the first time I'd ever been allowed at the mall without adults, and I had my own money to spend. I took full advantage. I bought enough posters to paper my bedroom walls, and filled my closet with the latest shoes and the cutest tops. And I bought a pair of sexy low-rise jeans that fit me like blue denim skin. I knew Mami would not approve of them, so I kept them hidden, planning how to spring them on her.

My scheming for a quince was supplanted by scheming for the dream jeans. The jeans were in hand, too, so this was no hopeless dream.

It took a lot of teenage finesse to get Mami to let me wear them. She eventually got over the shock of that short three-inch-long zipper and that the snug denim covered only the crucial parts of my backside. Those jeans were so perfect, so cool, that I wore them to shreds. I embroidered over worn holes, added colorful patches, and finally, after years of wear, cut them off into shorts.

I still have those shorts. My daughter wore them this summer, and I see why my mother objected. Man, are they cut low. I quickly took them back, claiming that they're an important reminder of my fifteenth birthday.

I enjoyed myself for many years in those great pair of jeans that came to represent my quince.

No lurid party tales.

Not that year, anyway.

Recuerdo: My Sister Remembered

BY Erasmo
Guerra

*I*n the picture, my sister, Michelle, sits in her wrought-iron
throne, and her long white gown drops to the wood floor of
the rental hall. It is her fifteenth birthday. She's a *señorita*
now. In my mind though, she will forever remain a teenage princess, dead before she was able to fulfill her promise as a woman in
the world. She holds a bouquet of pink ribbon and carnations. My
mother wears a matching corsage pinned to her rust-colored dress.
My kid brother, Marco, stands taller than most of us, wearing

slacks, a button-down shirt, and cowboy boots in dusty shades of brown. My baggy shirt and pants shimmer with the iridescence of middle-child neglect. Behind me, forced into a pale gray suit, my father stands with his shirt open at the collar. His mustache is dark. His hair brushed back. He faces the camera with his eyes closed, as if he wants to forget this night, or the more terrible night that will come later and stamp a permanent shadow over this family. For now we are caught in the glare of a photographer's flashbulb.

Many years later, while I am home on a visit, I sit in the gloom of the dining room of my parents' new house. The shades are drawn against the hot summer afternoon. This is the Valley, where most of my family still lives, and where the sun shines more than three hundred days out of the year. I moved away years ago and only come back for the memories.

I flip through the photo album the hired photographer had prepared for my sister's quinceañera. My father, in his sixties, his hair and mustache now gray, looks at himself in photo after photo and doesn't know what became of that suit. He scratches at the collar of his V-neck undershirt, at the tattoos that have turned green, remembering a much earlier night as a teenager, when he escorted a quinceañera.

"One of my ex-girlfriends," he says. He laughs and then mutters that he shouldn't say more because "your mom gets mad."

My mother, who used to pay me a nickel when I was a kid for each white hair I plucked from her head, would be completely gray right now if it weren't for the dye she works into her hair every couple months. Overhearing us, she comes in from another part of the house and asks why she would get mad. I tell her. She shakes her head and sighs, "*Déjalo que diga.*"

Not like my father remembers that well. He can't even recall his girlfriend's first name. Just the last—Martinez. "We were going steady," he says. He was either sixteen or seventeen or eighteen. He's not sure. His girlfriend was younger, but then, the way he tells it, "the dreaded moment came." Martinez turned fifteen and needed a *chambelán* to escort her at her quinceañera.

He remembers that instead of a tuxedo he wore a "regular suit."

"I don't know from where," my mother cracks. She's always considered him a horrible dresser, the kind of guy who wears white athletic socks with dress shoes.

"I dunno if I borrowed it or what, *pero* I had a suit," my father insists. "It was kind of a bluish suit."

My father says he really didn't want to go through with it. "I was from the *rancho*," he says, meaning that he didn't grow up in town like his girlfriend, but in the outlying community of Madero, where recent arrivals from across the river settled and raised families and worked at the German-owned brick factory nearby, which they called the Little Prison because it was hard work. On weekends, in your neighbor's backyard, cow heads slow-cooked in ember-filled holes in the ground. My father, who grew up there and then along the murky banks of the Río Grande, says he doesn't remember any of the local girls holding a quinceañera. His sister, Lucy, never had one.

"It was a city thing," he says, and then groans, "I was not a good dancer. That was the worst part—I had to be the first one dancing."

He lets out a low whistle as he tries to remember where the reception was held. It may have been at Mitla Patio, the south side hall where most of the weekend dances were held at that time, but he's not certain. "No, no, I think it was at the convention center." Only there is no convention center. He must mean the community center at Lion's Park. "Yeah," he seems sure of himself now, "I think it was the community center."

Then he remembers the girl's full name: Norma Linda Martinez, but he still can't recall where the reception had been held. Maybe it was at the Mitla after all. The lights strung overhead. The bandstand which he used to crawl under when he was younger to watch the folks dance and, the empty lot in the back where he and his friends played *canicas*—games of marbles with names like *el hoga'o* and *la chusa*. He knew more about the strategies and rules for shooting marbles than Norma Linda Martinez. When I press my father for more memories he says he doesn't have any. "Nothing," he insists. "*Na'a.*"

It's not so easy forgetting the night of my sister's quinceañera. A framed picture stands on the living-room coffee table. He sees it every day—my sister posed between him and my mother— and he says he feels numb. "I don't think. I don't wish. I don't— nothing. I just look at it."

My mother reminds me that I was a *chambelán* once. I'm not sure what year that was, how old I may have been, but my mother says, "*Allá 'sta el retrato,*" as if the picture will do the hard work of remembering.

All I recall is the afternoon I raced from a friend's house on my ten-speed so I could get home and get dressed in time to make the reception. Finally, a date with Iris, my junior high school crush, though we both were in high school by then. I shot through an intersection, under a traffic light turning red, and when I next looked up to see what was ahead, I saw the grille of an oncoming diesel truck and swerved left. Then nothing.

The emergency room doctor said I was fine. So I went home and put on my rented tuxedo, snapped on the aquamarine bow tie and cummerbund, passed a cloth over my spit-shined shoes, and headed to the dance.

There are pictures of me wearing a white boutonniere in my buttonhole. My date was cinched into a tight cocktail dress layered with lace, white gloves, pumps, and a veiled pillbox hat.

Otherwise I remember nothing. I blame the concussion.

My mother spent most of her life stooped over a sun-beaten cotton row and, at the end of a blistering work day that began at dawn and ended at dusk, went home to a tar-papered house where the drinking water was drawn in buckets from the irrigation canals. She was a champion picker who harvested a thousand pounds of cotton in a single day. Her picture appeared in the local papers. Just never to announce the occasion of her quinceañera.

"I don't even remember when I was fifteen," she says. "I never took my years seriously, like saying, 'Hey, tomorrow's my birthday,' *no le ponía cuida'o.*"

But she remembers the first time a guy came to her house to call on her. "I still played with dolls *en ese tiempo y yo lloré cuando me nombró.*" She says a sister-in-law told her, "*Oye*, you're already a *señorita*, it's time for men to come calling." But my mother cried no. "*Todavía no*—I still considered myself a little girl."

My mother would've been fifteen in 1959, but she doesn't recall anything in particular about that year. "*Como te digo, no me recuerdo nada porque* most of the time I was always working. I had to drop outta school to work in the fields. To help out. Mom was sick of nerves. Sick of her heart. So *que servia que supiera* that it was my birthday. There wasn't gonna be any gifts. I wasn't gonna have a party. I just lived my life working in the fields and taking care of Mom and that was it. I was happy." She worked hard and her only pleasures consisted of swimming in the canals, running through sprinklers, and standing in the rain. My mother always thought she'd make a great mermaid.

Still, she claims she was a *dama* for many quinceañeras, and she explains that she would buy fabric at the downtown dry goods store and then take it to the appointed seamstress, who made the dresses for the entire court of *damas.* "At that time it was cheap. It's not as bad as it is now. *Ahora te sale bien caro. Bien* 'spensive. A lot of people don't go with the *damas* anymore because a lot of them don't have the money, or the hassle *con la costurera, que los vestidos no quedaron bien. Y luego despues los chambelanes no todos quieren pagar.*"

Unlike my father, who wore that bluish suit for his *chambelán* duties, my mother says all of her escorts wore tuxedos. She recalls lining up with her date and standing under a flower-decorated arch as the *damas* and *chambelanes* were called. The way she tells it, the presentation started with a one-year-old *dama* and her escort (I don't know whether to believe her, but she goes on) then the two-year-olds, and so on. "*Las* girls *p'acá y los* boys *p'allá—hasta la que tenía* fourteen. *Y luego al último, la quinceañera sola con su chambelán. Bailaban el primer vals. Bailaba con el papá. Y luego al rato bailábamos todos.*"

My sister's quinceañera was my mother's idea. Maybe, since my sister was born on her birthday, December 28, my mother also wanted to celebrate her fifteenth, twenty-four years late. So in less than a month with a thousand dollars to spend, my mother and sister put the event together. They set the date for the last day of the year, the eve of 1984.

A local printer designed the invitations in Spanish, which none of us kids could read well. And the picture of my sister they used for the front of the invite turned out looking like an amateur photocopy job of gray tones and streaks.

Tía Ofelia lent my sister the three-hundred-dollar dress our cousin had worn a year earlier. "Just take it to the cleaners," Ofelia said.

On the big night, my mother helped my sister into her white dress—silk bodice, scalloped neckline, and puffy shoulders that tapered to the wrists. My mother had heard the superstition that *"las perlas eran malas,"* that they were bad luck, that whoever wore them would end up crying later, but she didn't want to believe it as she clasped the pearl necklace around my sister's neck and then adjusted her matching pearl earrings.

A family friend picked my sister up at our old house in a burgundy Lincoln Continental decorated with pink streamers and a quinceañera doll lashed to the front hood ornament. My sister, sitting alone in the backseat, rode out of that *colonia* in which we lived. Past the wood homes squatting on crumbling bricks. Homes without proper water or sewer lines so that a stink rose whenever the backyard septic tanks overflowed. Homes that were boarded up in the summer when families went north, following the harvest seasons for strawberries and corn. Homes where the neighborhood girls called one another stuck up for thinking they were better than this. Homes where the dogs prowled about glistening from baths of burned motor oil that was supposed to cure them of the mange.

The Lincoln Continental left all this behind as it made its way around the potholes and headed toward town to the early-evening Mass at St. Paul's. From outside, the concrete church, with its high arched ceiling, has always looked like a holy airport hangar. Inside, a huge copper and brass crown floated over the altar. My kid brother fantasized, as he did most Sundays during Mass, "What if the crown fell on the priest?"

My mother and father escorted my sister down the aisle to the altar railing, where she knelt on a fringed pink pillow embroidered with the sentiment MIS XV AÑOS. A Christmas tree stood decorated with gold ornaments and white doves.

The first three pews had been decorated with flowers, reserved for family, but, since this was a regular Saturday-evening Mass, most of the church was flocked by "Snow Birds"—white-haired retirees from the Midwest who migrated south to play shuffleboard through the warm winter. "*Se miraban asustados,*" noted my mother. "Like they'd never seen a quinceañera before."

The reception was held at the church hall next door, which we decorated with pink streamers, and filled with the melodies of "Happy Birthday" and "*Las Mañanitas.*" We ate on the butcher-paper-covered folding tables.

A few families pitched in with side dishes of Spanish rice and beans *a la chara* that smelled of bacon fat and cilantro and garlic, to ease the gas. Our family provided the main course of barbecue brisket. My mother doesn't know where she bought it. "*La mera verdad no me 'cuerdo—pero* I think I bought it *en* McAllen or Mission." She thinks a moment. "Ramiro? Ramiro's Meat Market?" She's not sure. "*Allá pa' rumbo* Palm View. The man already died—diabetes. They chopped off one leg. Then they chopped off the other. Then his wife died of stomach cancer."

The cake was baked in Edinburg. My mother doesn't know where, though she recalls that it was three tiers tall, the layers filled with pineapple. The *pan de polvo* came from the bakery at the H.E.B. supermarket where my mother worked.

The Barbie doll *muñeca* that sat near the cake came from the widow who lived in the *colonia* with her unmarried daughters. There was also a guest book RECUERDOS DE MIS XV AÑOS, but nobody signed it because we forgot to bring it out.

My sister posed for photos: Michelle with her lace parasol. Michelle with her school friend Maritza, who was the only friend she invited, because, as my mother had to explain, we couldn't afford to entertain everyone. Michelle with family and family friends—Chapa, Garza, Guerra, Ortiz, Perez, Riojas, and Tanguma. The women wore borrowed dresses and blouses and sucked in their stomachs so that they wouldn't look too fat. The men wore stretchy dress pants and boots with matching cowboy hats, smiling as if they wanted to let loose but wouldn't.

My sister had long black hair, which at times she blow-dried straight or burned into tight curls with a hot iron as she did that night. When my mother had the money, and even when she didn't, she would take my sister to the M&H Beauty Salon, with its narcotic smell of hair dyes and styling products, and my brother and I would sit on the floor by the old-lady beehive hair dryers, watching my sister get a spiral perm.

When girls sprayed their bangs into a high bird's crest, my sister would use an entire can of Aqua Net and then walk to the bus stop facing the wind's direction. She walked without seeing where she was going, her neck turning like a weather vane, so that her hair wouldn't get blown out of place. She was always a stumble away from disaster.

Her legs bowed like our 'uela Torina's. And she was obsessed with a "bump" on the bridge of her nose, which she swore she'd fix as soon as she got famous. A wart grew on the inside of one nostril, perhaps from obsessing so much, and my brother and I watched the family doctor cauterize the monster. She also suffered from ingrown hairs under her arm, and she cried from the pain and claimed that they were tumors. My mother put warm compresses on the pimples and, with a sterilized needle, lanced and drained them and told her no, she was not dying. And yet she looked so alive in those quinceañera pictures.

For her big night, my sister didn't have a live band. No mariachi. No DJ. Instead, we spun our own records: Michael Jackson's *Off the Wall*, the *Flashdance* soundtrack, and Laura Branigan with her ballad of grief "How Am I Supposed to Live Without You." They were the hot sounds that year.

My mother sensed that my sister would have liked a formal

dance—unless it was her own dancing feet trying to convince her—but my father didn't want anything. "*Tu papá así es—bien avergüenzoso*," my mother gripes more than two decades later. "*El no quería na'a*, period, like always, he never wants nothing!"

Not like they did anything for my birthday earlier that same year, when I finally became a teenager, going from twelve to thirteen. I waited for someone in my family to say, "Happy Birthday." My parents didn't even remember that it was their wedding anniversary. (You'd think that my being born on their anniversary would help them remember, but maybe they just wanted to forget.)

I sat at the kitchen table all night. As everyone went off to bed, my mother turned out the kitchen light and told me, "*Ya, duérmete.*" I didn't move. She asked, "*¿Por qué te pasa, huerco?*" I started to cry in the dark, letting out these very ungrownup gasps, telling her that it was my birthday today.

Everyone got out of bed and climbed into the El Camino and we drove down the road to the convenience store. I think it was called the Sunshine. They were about to close, but the counter guy let us in and as everyone waited in the car I wandered the fluorescent-lit aisles, looking at the racks of fried pocket pies, and their stock of pre-packaged goods from the Butter Crust bakery upstate.

I picked out a German chocolate cake wrapped in cellophane and blue plaid cardboard. It wasn't even the whole cake. Only half. They sold it in half and quarter portions. For a long time afterward I loved German chocolate cake. Now I can't stand it.

My parents went on forgetting my birthday. A kid as ignored as me would have run away from home a long time ago but I guess I was fine with it. I never wanted the limelight as much as my sister, who warned our parents that as soon as she graduated from high school, she was leaving.

We lived in Mission, a town of twenty-five thousand, famous for being home of the Ruby Red grapefruit and the birthplace of Dallas Cowboy coach Tom Landry. My sister wanted a bigger and more glamorous life away from all that. My brother and I didn't know what she had to complain about since she was the favorite. She got the latest Gloria Vanderbilts and Sergio Valentes. So what if they were from K-Mart. My sister even had her own room— well, we did too but it just wasn't the same—with a canopy bed, pre-op Michael Jackson posters, and walls the color of pink flamingos on fire.

In high school, she danced with the High Flyers, the majorettes who dressed as cowgirls with tasseled white boots, ruffled skirts, vests, and gloves. At the Friday-night football games, they tipped their cowboy hats at the start of every half-time show. My sister languished among their lower ranks for two years before she made the more respected cheerleading squad. She hollered and shook her maroon pom-poms at the games. If our team, the Eagles, wasn't doing well, the cheerleaders snapped out their double-jointed arms and stiffened their hands into talons and hissed the battle cry, "Eagle Claw." The other popular cheer was called "Go Fight Win," which they did to the accompaniment of the marching band:

GO. GO-GO. GO, MIGHTY EAGLES.
FIGHT. FIGHT-FIGHT. FIGHT, MIGHTY EAGLES.
WIN. WIN-WIN. WIN, MIGHTY EAGLES.
GO, GO, GO—FIGHT, FIGHT, FIGHT—WIN!

She was a spirited high school student who was junior class president, head of the prom committee, and who worked on the school paper and yearbook. After school, she worked as a cashier at the Mission H.E.B. supermarket. On weekends, she volunteered at the Retama Nursing Home in McAllen. Like most teenagers, my sister was busy with life. And she was my sister. Diana Michelle Guerra.

The last time I saw her alive, she was sitting on my twin bed, in the yellowish light of my room, watching me pack a suitcase. It was the summer of '87. The radio wept with ballads of displacement like "Somewhere Out There," by Ronstadt and Ingram, or it rocked with the power chords of "Alone" from the sister duo Heart. I was seventeen and going away to Florida to stay with cousins. I planned to get a summer job at Disney World.

My sister, eighteen, had just graduated from high school. Her plan was to go to North Texas State in Denton, to study I don't know what, since she was always changing her mind. She wanted to be a veterinarian, a high-fashion model in Paris or Milan, or an actress on *General Hospital* or *Guiding Light*. She had the dizzy and undisciplined aspirations that everyone has in their youth. She wanted to get away. She howled her entire short life that she couldn't wait to escape.

Our parents wanted her to stay put and attend Pan American (which my friends and I called "Taco Tech"), because it was less than ten miles from our house, and because she was a girl. She would be safer here.

My sister had not been anywhere on her own other than the cheerleading camps in San Antonio and Dallas, which had been chaperoned by squad sponsors. There was a Spring Break trip her senior year to Puerto Vallarta, but that had been with my mother, and the report back was that they had fought. One of my sister's favorite arguments, as she got older and demanded she be allowed to go where she wanted, was to say, "If I'm gonna die, I'm gonna die." Her teenage fatalism never convinced our parents. More than anything, they were shocked by her desperate, almost suicidal challenge to be let out into the world.

In July I bought a postcard and wrote my sister, encouraging her to come to Florida to audition for one of the Disney shows. But I never got the chance to mail it. The night of the fourth, returning from my job at the Pecos Bill Café in Frontierland, I felt homesick and alone. My cousins were at a barbecue. I stayed in, feeling sorry for myself, which my sister would not have approved of since she was forever advising me that I needed more friends.

I spent the night switching television channels and felt a strange thing happening to me that only later I identified as growing up. Only it wasn't the sudden social flowering that a woman experiences at fifteen, coming into her own with the blessings of the church, her family, her friends. I, like most men, did it alone and without ceremony. I decided that night that if this was adulthood, I wanted to take the next flight back.

Later, as I tried to sleep, the neighbors fired guns into the night and set off fireworks in celebration of our independence. A dog howled as I recalled past July Fourths when my brother and sister and I made our parents buy us bottle rockets and smoke bombs. We lit sparklers at the kitchen stove and ran through the hall and out the door as my mother screamed about the house catching fire. When my parents fell asleep, we snuck into the fridge and took a can of my father's Lone Star and passed it around.

Early the following morning my cousin knocked on my bedroom door and said my mother was on the phone. Then I knew. The dog from last night. Mexicans believe the superstition that dogs howl when someone you know has died. When I got to the phone, all I heard was my mother's choked crying.

"They took her," she said.

* * *

I flew home the next day. When I arrived, the sheriff's department had questioned everyone from family to neighbors to current and past boyfriends to suspicious uncles who they felt were too interested in what had been found so far. Which was nothing. She was still missing.

I drove my father's truck into town and put up flyers with pictures of my sister and the question: HAVE YOU SEEN HER?

Later, I rode to a house where a group of my sister's friends had gathered to pray, and I stood alone with my brother, as my sister's boyfriend and her best friend cried together in the middle of the living-room couch. When the crowd dispersed, everyone going home to dinner and the rest of their lives, a call came through from an uncle telling my brother and me to get home.

"Did you find her?" I said on the phone. "Is she all right?"

He refused to say anything specific, just that we should get home, now.

Outside, as we looked for a ride, a brown car came screeching down the street and almost came up onto the yard. The car rocked back and forth as it came to a stop. The girl, another of my sister's friends—she had so many—flung the door open and shouted that they had found my sister.

"They found Michelle," she cried, and before taking another breath, she said that my sister was dead.

My knees buckled. I fell onto the grass. White clouds drifted in the burning sky and I felt the heat taking me, taking me, taking me into its arms in a way that all of my sister's friends had failed to do.

They found my sister in a pasture behind our house. The summer heat had badly decomposed her body. The local news stations broadcast grainy video clips of her skinny cheerleader's arm jutting out of the tall dry grass. The phantom lights of the camera shone against skin that looked too pale to have been hers.

For someone who had such far-flung dreams—and this is the part that makes my heart ache most—her life came to this nightmare end not far from home, where she was supposed to be the safest. They never found her killer. The case went cold.

After high school, I moved away without actually moving on. My sister's friends attended colleges near and far; they married and tacked on their husband's last names; they became registered nurses and teachers and mothers. In my own family, the next generation of cousins continues to celebrate their passage into womanhood. They now spend up to a year and thousands of dollars to plan and pay for the event. They order their triple-tiered cakes and sugary *pan de polvo* cookies from the Celebrity bakery in McAllen. They buy their dresses and tiaras from the Princess shop in Palmview. They rent Nellie's Ballroom, the Villa Real Convention Center or the Outta Town Dance Hall. They pay up to two hundred dollars to print a two-hundred-word announcement in the local newspaper.

My second cousin Danielle says she was given the choice of a car or having a quinceañera. She chose the party. "My dress was made especially for me," she says. "It was drenched in pearls and sequins, white silk with off-the-shoulder straps." She wore matching heels, which got lost when she stashed them under the cake table, but the high point of the night was dancing with her father. The song she chose was "Hero" by Mariah Carey. It was especially memorable, Danielle says, because her father passed away a few short years later. "I only have what my father left me—memories to last a lifetime."

Another cousin, Melinda Yvette, says she didn't have the tradi-

tional court of *damas*, "because it ends up being a problem finding
the right dress to fit everyone. So, I went by myself." Like a bride
throwing her garter belt, Melinda Yvette tossed a doll to a group
of younger girls, symbolizing her "last doll." Years later, she insists that a quinceañera is "something that you'll never forget even
when you get older. I felt happy that day."

Yet another cousin, Cassandra, who celebrated the most recent
quinceañera in our family, describes hers this way:

> The photo people arrived at my home around 3 p.m. to videotape me
> and my family. Then, around 3:30, a limo excursion picked up my
> friends and my little brother and me and we rode around for two
> hours. Anywhere we wanted to go, we just picked up the phone and
> told the limo driver.
>
> At 5:30 he took us to the Seven Oaks Country Club. My mom had
> people decorate the place with beige table covers and each chair had a
> ribbon wrapped around it with a beautiful bow on the back. I had my
> own table for my friends. I had beautiful fresh lilac tulips and fish
> bowls with a beta fish in each. The fish were given away to my friends
> as a gift from me.
>
> I wore a long lilac dress with pearls and sequins and a small tiara
> with matching jewelry. My mom wore a long maroon beautiful dress
> and my dad and brother wore black pants and white shirts, not being
> tuxedo people.

Cassandra says she didn't have a dance with her dad because
she was too shy. Still, as shy as she may have been, for one night
she enjoyed being the center of attention and says she will forever
cherish the photos and video as a memory.

Dear Michelle, on my last trip back home, Mom, obsessed
with graveyard real estate, takes me on a drive-through of Valley

Memorial Gardens. She points out the new section where she has moved herself and Dad, because she didn't want to be too far from you and the two plots on either side that have been bought for Marco and me to escort you into eternal rest.

Mom, once busy making a perfect home, now prepares us for the afterlife. She insists, "*La muerte de segura la tiene uno. Por mas que uno no quiere hablar de la muerte*, you have to know that everybody someday will have to leave. Nobody is gonna stay. *Un día como quiera te tienes que ir al otro mundo.*"

The sprinklers shoot weepy arcs of water over the grounds. We don't get out of the truck to see you. Besides, from the countless trips I've made to your grave, I have memorized the inscription on the marker that reads "She shared her smiles and hid her tears." Mom kisses her fingers and waves them toward you and whispers, "Love you, *mija.*"

She points out the nearby tree that was planted around the time of your burial. It now casts a shadow over your grave in the afternoon. "She don't have to be in the sun that much," Mom murmurs, forgetting that you always wanted to be in the sun. To shine under a Hollywood spotlight.

Pointing out the distance between the graves where you and Marco and I will eventually lay, and where she and Dad will be buried, Mom says she didn't want us ending up too far away from one another. She wants to keep us together forever. She wonders aloud, "You think I did right by changing?"

She's still not sure. But one thing is certain. This is the end of the line for us Guerras. Marco and I are both unmarried and childless. There's no one to carry on our name. Soon, instead of a family remembered, we will be a family forgotten. "Remember *los* Guerras," they'll say, and no one will.

Later, Mom drives to the old church, where you took your First Communion, celebrated your quinceañera, and where we held your funeral services. As we get off the truck, Mom grumbles that she should have brought her hat. She puts a folded newspaper over her head instead. "I gotta protect myself," she says, concerned ever since the skin cancer diagnosis a few years ago. "I'm not gonna let the sun get to me."

We try the heavy doors to both the church and the new parish hall—the old one was demolished—but they're locked. We're not sure where to go next. What doors of memory are left?

The palms sway in the summer breeze and the *anacuahita* trees drop their white trumpet flowers that burn on the hot sidewalk. A flock of wild parrots bursts into the sky, their green feathers glimmering in the sun as they swoop and squawk. I never knew parrots roosted in the Valley, but Mom says, "They come from Mexico— they cross."

I want to think there are fifteen altogether, but I lose count after twelve. By Mom's estimate they are merely a bunch. "Look," she whispers to herself, "there they go." The birds flutter past in an emerald blur, between the tall palms and over the nearby park, the flock unraveling and disappearing into the sky.

Contributors

Alberto Rosas was born in California and has acted in a few plays. Everyone he met said, "My heart is in the theatre." Alberto hoped that each play would bring him closer to finding his heart. After appearing in about seven plays, Alberto still hadn't found his passion. His girlfriend at the time said he was heartless. Though he disagreed, Alberto nevertheless quit the theatre. In addition to various short films and television appearances, Alberto starred in five feature-length independent films, portraying such different characters as a juvenile delinquent, a heartbroken homosexual, a man accused of child molestation, a drug kingpin, and a poet-writer. Alberto currently attends law school. He admits that law school is a lot like working with actors. In his spare time, he is completing work on a crime novel. Occasionally someone still tells him that he looks like Antonio Banderas. Alberto no longer attends quinceañeras. For more information about Alberto's writing and film work, please visit www.albertorosas.net.

Angie Cruz was conceived in the Dominican Republic and born in New York City's Washington Heights. She went to La Guardia High School concentrating on Visual Arts and followed that path to study Fashion Design at F.I.T. In 1994 she resigned from her fashionista lifestyle to become a full-time college student at SUNY Binghamton, where her love affair with literature began. She graduated from the NYU MFA program in 1999. Cruz has contributed shorter works to numerous periodicals including the *New York Times*, *Latina* magazine and *Callaloo*. She published two novels, *Soledad* and *Let It Rain Coffee* (Simon & Schuster). She currently is working on the screenplay for *Soledad*, optioned by Nueva York Productions. She recently transplanted herself to Tejas to work as an assistant professor at Texas A&M, where she is finishing her third novel.

Constanza Jaramillo-Cathcart was born in Medellín, Colombia, in 1972. She studied Comparative Literature and Film and has a masters degree in journalism. She has lived in Colombia, Mexico, France, and the U.S. She currently lives in Brooklyn, NY, with her husband, Blake, and her son, Lucas, and works as a teacher and writer. Her fiction has appeared in the *Brooklyn Rail* magazine, the *Brooklyn Rail Anthology*, and she is a member of LART (Latino Artists Roundtable). She's at work on her first novel, *Subtítulos para la vida* (*Subtitles for Life*).

Fabiola Santiago is a features writer at *The Miami Herald*, where she has been a journalist since 1980. Her articles on Cuban culture, arts, and identity have been published in many U.S. newspapers and magazines, and in Latin America, Canada, and France. Her poetry and short stories have appeared in *The Caribbean Writer*, *Tropic Magazine*, and *Highlights for Children*. Born in Matanzas, Cuba, she came to the United States with her family in 1969 when she was 10 years old. She lives in Miami.

Leila Cobo-Hanlon is the Bureau Chief for *Billboard* magazine's Latin/Miami editions. Her work has appeared in *Latina*, *The Miami Herald*, and several other publications. She was born in Baranquilla, Colombia, and lives in Key Biscayne. She is at work on her first novel.

Nanette Guadiano-Campos is a writer/teacher and *mexicana* hailing from the Lone Star State. During the day, she teaches children how to read and write. At night, she reads and writes. She has had several poems published in *Border Senses, Flashquake, True Poet Magazine* (Top Ten Choice, June 2006), *Mom Writer's Literary Magazine, The San Antonio Express News*, and other e-zines. She just placed first, second, and honorable mention in the San Antonio Poetry Fair and will be featured in an anthology entitled *Voices Along the River* later this year. Though Nanette loves writing poetry, her first love is prose, and this essay is her first published piece. She is currently working on a book of short stories based on her Mexican-American heritage, as well as polishing up her first novel. She lives in San Antonio, Texas, with her husband and two daughters.

Malín Alegría-Ramírez was raised in San Francisco's Mission District. She is a graduate of UC Santa Barbara and received her MA in Education. She is a teacher, eco-warrior, Aztec dancer, and performer. She has written various short stories, performances, and has worked with groups like Teatro Nopal, Lovefest, and the WILL Collective. Her first novel, *Estrella's Quinceañera*, was released in 2006 by Simon & Schuster. Malín currently lives in California and is at work on her second novel, due out in March 2007.

Adelina Anthony is a self-identified genre-whore, meaning she does it all and doesn't feel guilty. A performer, writer, director, and digital media artist, this scandalous Chicana lesbian just graduated with her MA from Stanford, where she studied intensely with her mentor, Cherríe Moraga. She is a former PEN USA Rosenthal fellow. Adelina has been featured in *Nerve, Best American Erotica 2002, Bedroom Eyes, Tongues*, and other zines. Her play,

Mastering Sex & Tortillas, is currently being considered for publication.

Eric Taylor-Aragón was born in Berkeley, California, in 1972. His mother is from Peru and his father is from England. He has an undergraduate degree in interdisciplinary studies from UC Berkeley, where his studies were focused on Modern Literature and Philosophy. He has lived extensively in Latin America, where he did human rights work after university. He has been published in Ishmael Reed's *Conch* magazine and in the New York–based literary journal *Fence*. He has contributed as a researcher/writer on two bestselling nonfiction books, *Savages* by Joe Kane and *Ptown—Art, Sex, and Money on the Outer Cape* by Peter Manso. Aragón is currently shopping his first novel and finishing up a collection of short stories.

Barbara Ferrer is a first-generation, bilingual Cuban-American, born in Manhattan and raised in Miami, all of which she realizes makes her a walking cliché. However, it also means she has at her disposal a large, extensive family from which to draw inspiration and an arsenal of colorful expressions to use in her writing. Writing as Caridad Ferrer, *Adiós to My Old Life* was released by MTV Books in 2006, garnering praise such as "A page-turning must-read" (*Curled Up with a Good Kids' Book*), and ". . . an intelligent debut novel about the world of music and reality television" (*Romance Junkies*). Her second novel for MTV Books will be released in August 2007. Barb can be found on the Web at: http://barbaraferrer.com or www.caridadferrer.com.

Michael Jaime Becerra was raised in El Monte, California. He is the author of the short story collection *Every Night Is Ladies' Night* (HarperCollins, 2004). He is at work on a novel.

For 25 years, *Monica Palacios* has been on the forefront of queer Latina/o writing and performing. Creator of several one-person shows including *Get Your Feet Wet* and *Greetings From a Queer Señorita*. She has been anthologized in *Latinas On Stage*; *Out Of The Fringe: Latina/o Theatre & Performance*; *LA Gay & Lesbian Latino Arts Anthology 1988–2000*; *Puro Teatro*; *Living Chicana Theory*; *A Funny Time to Be Gay*; *Latina: Women's Voices from the Borderlands*; and *Chicana Lesbians: The Girls Our Mothers Warned Us About*. Her work continues to be studied at universities, and she is a lecturer at Loyola Marymount University, UCLA, American Academy of Dramatic Arts, UC Santa Barbara, California State University Los Angeles, and UC Riverside. Among her awards are the Postdoctoral Rockefeller Fellowship; Playwright Finalist Chicago Dramatists; Excellence in Playwriting—Indie Award; *OUT* Magazine's "OUT 100"; One of the Most Influential LGBT Latinas/os in the Country; and Los Angeles Latin Pride Foundation. She has also been commissioned by the Center Theatre Group. Find her on the Web at www.monicapalacios.com.

Felicia Luna Lemus is the author of two novels, *Like Son* (Akashic Books, 2007) and *Trace Elements of Random Tea Parties* (Farrar, Straus and Giroux, 2003). Her writing has appeared in *Bomb*, *Small Spiral Notebook*, *A Fictional History of the United States with Huge Chunks Missing*, *Latina* magazine, and elsewhere. She lives in New York City.

Berta Platas, the author of *Cinderella Lopez* (St. Martin's Press, 2006), began her writing career with Kensington's Encanto line. She was born in Cuba, raised in the United States, and now lives in Atlanta, Georgia. Her next book will be a novella in the anthology *Names I Call My Sister*, coming from Avon in 2007.

Erasmo Guerra was born and raised in the Rio Grande Valley of South Texas. He is the author of the novel *Between Dances*, which won the Lambda Literary Award, and he is the editor of the nonfiction collection *Latin Lovers: True Stories of Latin Men in Love*. His work has appeared in a number of journals, magazines, newspapers, and anthologies, including *New World: Young Latino Writers* and *Hecho en Tejas*. Guerra has been awarded grants from the Vermont Studio Center, the Fine Arts Work Center in Provincetown, and the Virginia Center for the Creative Arts. He is a member of Macondo, the writing collective headed by Sandra Cisneros. He lives in New York City.

Acknowledgments

Thank you: John Hughes, Molly Ringwald, and Anthony Michael Hall for guiding us through the weird times. To René Alegría, for the opportunity to feel like Samantha from *Sixteen Candles* rather than Hope from *Thirtysomething*. *Gracias* to Melinda Moore and the Rayo team for making the work plain fun. Much thanks is also due to my agent Joy Tutela as well as Michelle Herrera-Mulligan, Marcela Landres, Carmen Ospina, Deborah Kreisman-Title, and Rafael López, for their input and support.